SHORT Stories

by Ralph L Myers

Printed in the United States of America

ISBN 979-8-89114-313-5 (sc)
ISBN 979-8-89114-277-0 (e)

Library of Congress Control Number: 2026906147

2026.07.01

MainSpring Books
5901 W. Century Blvd
Suite 750
Los Angeles, CA, US, 90045

www.mainspringbooks.com

Contents

The Biloxi Kid – Confederate Cowboy

A Short Story
By Ralph L Myers

The Biloxi Kid — Confederate Cowboy

Jebediah D'Frain, a captain of the Confederate States Army, shakes the dust and loose grime from the remnants of his Confederate uniform as he dismounts his black and white stallion at the edge of a rapidly running stream. "Better fill up, girl; it might be another day before we will find another water stream of cool, clean water like this one here," stroking her mane, which appeared gray because of the accumulation of dust embedded in it. Taking the army-issued canteen from the saddle, he gulped down what little was left and then bent down at the bank of the stream next to his mare as she noisily lapped up as much of the water she could, submerging it in the crisp, clean water to fill to the top.

Spotting a group of trees just on the other side of the stream, he mutters, "Eh, it looks like a good place to set up camp for the night, doesn't it, old gal?" His mare, still drinking her fill of the water, shook her head up and down and whinnied as though she was answering him, and then quickly returned to lapping up more water.

Sitting down on the bank of the stream and leaning back against a large tree stump of a fallen oak, he stretches out his 6'3" frame and pulls out a tattered, folded piece of paper that was tearing where it was folded. Carefully unfolding it, he looked at a hand-drawn map of what was supposed to outline the states of Mississippi, Louisiana, and Texas. *"Let's see, it's been over three weeks since I left Biloxi, so if I have averaged 60 to 100 miles a day, I should get within a day's ride of the Louisiana, Texas border,"* he thought. Turning to his horse, he tells her, "In about another week, we should arrive in Sage Weed, Texas, old gal. There, we will meet up with

Colonel Bartson and his troops. Think you can make it?" This time, he was ignored by the mare that was now grazing on the grass along the bank of the stream.

Captain Jebediah D'Frain, the only child of French Cajun parents who had settled in Mississippi after moving from Baton Rouge, Louisiana, was nearing his 23rd birthday. Knowing that war would eventually break out between the North and the South, his parents wanted to protect him from being taken by any of the ragtag, disorganized bands of renegades and forced to fight in their groups, so they enrolled him in a military academy for teens when he was fifteen years old. By eighteen, he had graduated from school, reported directly to Confederate Army troops commanded by General Lee, and was awarded a commission as a lieutenant.

While attending the military academy, he became known as Jeb, the kid from Biloxi, Mississippi, a name shortened by his instructors and classmates to the Biloxi Kid. This name had followed him ever since. Even after he was promoted to captain and given command of his own company of troops, the soldiers that served under him referred to him (privately amongst themselves) as the Biloxi Kid, as did some of his superior officers in the regiment.

The first strains of daylight were breaking through the trees along the bank of the stream where the Biloxi Kid and his mare had camped overnight. The plaintive howls of a lone wolf could be heard off in the distance as the darkness of the moonlit night was rapidly fading into the first rays of the early morning sun. Digging with his trenching tool, he quickly extinguished the barely glowing wood in the campfire with dirt. Rolling up his poncho, which had served as a sleeping bag, he walked over to where he had tied his mare, picking up the saddle and throwing it on her back. "Time to leave, old gal," he softly tells the horse we have got a lot of territory to cover before our journey ends. "Besides, we have to keep a lot of miles between Johnny Spencer and his bunch of bounty hunters who want to take us back to Tennessee, dead or alive; it sure as heck doesn't matter to them because they will get the same amount of blood money either way. "In fact, I would bet you a bag of oats, old gal. They would probably prefer dead rather than living with me tied over your back, instead of being able to cause them trouble."

After securing the saddle to the mare and loading up his gear, Jeb fills up his canteen and the canvas bag with clean water from the stream and puts them on the horse as well. Taking off his clothing and going back to the stream, he wades up to his waist and then immerses his entire body, and then emerges to the surface, rubbing away all the dirt that was on him. Getting out of the stream, he towels himself off with the cleanest thing he had, one of the mare's blankets. Dressing quickly, he mounts the mare, his boots in the stirrups, digging into her side, yelling, "Giddy up, old gal, let's put Mississippi further behind us, Texas, here we come."

Bending over and running a piece of a tree branch through the dirt, uncovering the ashes from a campfire, the Indian guide traveling with Johnny Spencer tells him, "We're still on his trail, Johnny One Leg. He can't be more than a day or so ahead of us."

Getting off his horse and walking stiffly, supported by his good and his wooden leg, over to where the Indian was now standing, Johnny Spencer pulls out his Colt 45 revolver, "If you call me Johnny One Leg just one more time, I swear I'll empty my six gun in you, Indian guide or not. "You got that straight injun?"

The rest of the men remained on their horses, some of them laughing at the exchange that was taking place between Johnny Spencer and the Indian guide. Quickly, he became silent as he fired off two shots in their direction, knocking the hat off one of them. "If I hear one more laugh, the next bullet will hit somebody's head, not just his hat. "This hombre we are chasing I want bad, because this is personal, very personal, he and his rebel troops caused me to lose my leg during the battle at Gettysburg, and I am going to return the favor once we capture him, then I'm going to put a bullet in him. "Water your horses and get some fresh water for yourselves, as we are going to move out as soon as you do. "Biloxi Kid, we're going to catch you, and we'll soon meet again, I promise!" he yells.

It was getting close to midnight as the Biloxi Kid slowed the mare down to a walk as he entered the town of Sage Weed, Texas. He noticed the welcoming sign, "Welcome to Sage Weed, population 534, 398 living and 136 living in Sage Weed Hill Cemetery." Just below the sign were two smaller arrow-shaped signs pointing toward and advising: Dallas 501 miles, Houston 300 miles. From the town's eastern city limits, it was dark and quiet, but he could hear music and loud voices coming from what must be the center of Sage Weed, mixed in with the sound of an occasional gunshot as someone shot their pistol in the air. "We made it, old gal, we're here, Sage Weed." The mare snorted, moving her head from side to side. "Don't worry, old gal, we'll find you a stable and get you a big bag of oats, my trusted amigo, but first I need to find a place to bunk out for the night, and at daybreak we'll ride out and meet with Colonel Bartson and his troops."

Picking up the speed of the mare's gait, the Biloxi Kid headed towards the center of town, to where the music was coming from. Getting closer, he slowed the mare's pace back down to a walk, cautiously looking from side to side at anyone walking or drunkenly staggering on the boardwalks or out in the street. The mare was getting more nervous as they got closer to the town's center, snorting more, occasionally whinnying, ready to take off in a panicked run. "Easy, old gal, we're okay. Just think about that big bag of oats."

Finally, they reached a brightly lit, smoke-filled, and noisy building. Looking at the sign over the door, I read: "Tuscaloosa Lucy's Saloon and Hotel." Tying the mare to a hitching post, the Biloxi Kid walks into the large, loud, and smoke-filled room. At the front, there was a stage where about ten dancing girls were performing with at least twenty tables of loud, drunken cowboys hooting and cat-calling the girls as they danced. In the middle of the saloon, at least another twenty tables were accommodating six men at each table playing poker. Just beyond those tables was a bar that was over fifty feet long, filled with cowboys sitting on barstools and standing near or behind them. Behind the bar, which three bartenders were attending, was an enormous mirror that extended the entire length of the massive bar. Painted across the bar were the words, "Welcome to Tuscaloosa Lucy's, where you can leave your thirst and your money behind."

Walking past the gamblers up to the bar, the Biloxi Kid motions to one bartender, "Bartender, a whisky, please."

Bringing the glass of whisky over, the bartender says, "That'll be 2 bits," he says, handing him the full glass.

A moment later, he feels a tap on his shoulder. Tensing up, he puts his right hand on his holster and slowly turns, and is face to face with Lucille Florington, better known as Tuscaloosa Lucy, a stunning auburn-haired and large-bosomed woman wearing a revealing low-cut gown. "Easy with the gun, cowboy. I can tell you are not from around these parts, as I don't remember the last time I heard a man say "please." "Normally, it's just what I want and gimme."

Relaxing his right hand and taking it away from his pistol, he smiles, "Sorry, miss, you're right, I'm not from these parts."

"Judging from your manners and that southern drawl, I would wager that you're from the confederacy, probably Alabama or Mississippi."

"You'd lose that wager on a technicality miss. I'm really from Baton Rouge, LA, but I spent most of my years as a kid in Mississippi, so I probably have more of a Mississippi accent, and judging from the accent you have, I would guess you are from Mississippi as well."

Smiling, reaching for, and shaking his hand, I'm Lucille Florentine, the owner of this establishment. The folks around here call me Tuscaloosa Lucy mainly because of the name of my saloon, but I was first given that nickname when I moved to Sage Weed from Tuscaloosa, Alabama, at the start of the Civil War. "What's your name, cowboy?"

Returning her smile, "I'm Jebediah D'Frain. My family calls me Jeb, but most people know me by the name I was given while attending school: Kid, or The Biloxi Kid, since I hailed from Biloxi, Mississippi. Then, lowering his voice, he said, "As one southerner to another, I am also a Captain in the Confederate Army."

"What should I call you, Jeb, Jebediah, Kid, or Biloxi Kid?" Lucy asks?

"It doesn't matter to me as I answer any of them, then changing the subject, he says, "I need a place to stay for the night, and do you have any vacancies in this hotel?"

"I'll check, but I am certain we can accommodate you. Would you be interested in sharing a room with one of my girls?"

Smiling, "That sounds mighty tempting, but I need a good night's sleep, a bath, and a soft bed to sleep on. "Also, I need a stable where I can board my horse and get her the large bucket of oats I promised her."

"Yes, we have a stable across the street from the hotel, and my, my, such devotion to your horse. Would you be so considerate to your wife?"

"If I had a wife and if she was as loyal and took care of me as my horse, you betcha I would be."

Motioning to the bartender, Lucy tells him, "Give me the key to room 10."

"Here you are, Miss Lucy," the bartender says, handing it to her.

Turning and giving it to the Biloxi Kid, she replies, "Here you are, Biloxi Kid, I will take care of getting you registered, your horse fed, and have a bath drawn."

"How much do I owe you for Tuscaloosa, Lucy?"

"Tonight, it's on the house, but if you change your mind and want some feminine company, it will cost you $2.00," she laughingly answers.

After a warm bath and shaving, the Biloxi kid falls into the soft feather bed and instantly falls into a deep sleep. The last thought on his mind was how close Johnny Spencer was to catching up to him. He might not have fallen to sleep so quickly if he knew that at this very moment, Johnny One Leg and his bounty hunters were less than a day's ride from Sage Weed.

The air was stifling hot and humid on this early day in May as Johnny Spencer and his gang of bounty hunters rode into Sage Weed. They were loud as they rode into town. Immediately they headed for Tuscaloosa Lucy's Saloon and Hotel to wash away their thirst and the dirt and grime from days of riding in pursuit of the Biloxi Kid and any other escaping Confederate soldiers who were eager to join Colonel Bartson's forces who were still fighting the Federal Army even though General Robert E. Lee and the Confederate Forces had surrendered to Union General Ulysses S. Grant a month earlier in Appomattox, Virginia.

Johnny Spencer was followed into the saloon by the 20 men of his bounty hunter group. Their entry caused the saloon to become quiet for a few moments as they walked towards the massive bar. "Whisky for all my men," Johnny shouted to the bartenders as he sat on a barstool with his wooden leg dangling over the side.

Lucy, wanting to keep order in the saloon, sent several of her dance hall girls over to keep their minds focused on them rather than causing any disturbances by the twenty cowboys. "Keep them entertained, girls, and there will be a $1.00 bonus for each of you keeping them busy."

After several drinks, the dance hall girls started leading the now intoxicated men from the bar up into private rooms to earn the bonus.

Johnny Spencer was still seated at the bar and was approached by Lucy. "Hello, cowboy, what brings you and your men to Sage Weed?"

Eyeing Lucy suspiciously, Johnny asked in a drunken and slurred speech, "Who the hell are you, lady, the whore?" and without stopping answered her question. "My name is Johnny Spencer, and my men and I are here to arrest and bring back any Confederate soldiers who are trying to escape into Texas and the New West Frontier. Then, pausing and clumsily pulling a worn and faded "Wanted" poster from his shirt pocket, he holds it up. "Recognize this man; do you know where I can find him?"

Lucy, maintaining the composure of a seasoned poker player who won more than they lost because of their ability to bluff other players, recognized him immediately as being the Biloxi Kid, whom she had just met the night before. "Sorry, Johnny, I've never seen him before."

"Well, if you do, I want you to tell me, I have a personal score to settle with him," he tells Lucy as he takes a hunting knife from his pistol belt and plunges it into his wooden leg that was dangling from the barstool. Twisting and pulling the knife that was embedded in his leg, he tells Lucy, "The Biloxi kid and his rebel soldiers caused me to lose this leg at Gettysburg, and I want to return the favor before I kill him."

"I can promise you, you will be the last person I will talk to about seeing the Biloxi Kid. I hope he has checked out of the hotel like he said he was going to do," Lucy thought.

Although Lucy was unaware whether the Biloxi Kid was still in her hotel, she was worrying needlessly, as at that very moment, Captain Jebediah D'Frain, the Biloxi Kid, was reporting in person to Confederate Colonel Bartson on the Mexico side of the Rio Grande.

Saluting, Colonel Bartson welcomed the Biloxi Kid. "I am happy to have an officer of your caliber in my command, Captain.

"Thank you, Colonel, and I, too, am pleased to serve under your command."

"We have been advised by our Mexican friends that there is a large contingent of Union Army soldiers bivouacked on the other side of the river waiting to launch an attack on us."

"Colonel, you should also be made aware that there are about twenty or more men, bounty hunters, led by Johnny Spencer, headed this way and may be in Sage Weed as we speak. "They are to join the Union Army forces, and their mission is to capture as many Confederate soldiers as they can before they can escape into the new western frontier. "I also have been advised that there is a bounty on my head and that Johnny Spencer wants to deal with me."

"I have heard about this. Only I thought it was a rumor, Captain D'Frain. "I want you to take a squad of my soldiers dressed in civilian clothing after dark tonight and find out about this bunch of bounty hunters and then report back to me by 0900 tomorrow morning."

"Yes, sir, Colonel, I will trust your officers who know the men in this regiment to assign me five of their best men to accompany me on this mission. "Tonight, we will have the advantage of a new moon, and it will be pitch black out when we cross the river and ride into Sage Weed. "If my hunch is right, we should be able to gather a lot of intelligence at Tuscaloosa Lucy's Saloon, because any man entering Sage Weed would make that his first stop."

"Good luck, Captain D'Frain. I will see you back here at the command post at 0900 tomorrow morning," the Colonel says as he salutes and dismisses the Biloxi Kid.

Just as he had predicted, it was a moonless night, and it was impossible to see even the man in front of you as the five Confederate soldiers and the Biloxi Kid crossed the Rio Grande River from Mexico into Texas. Finding a shallow crossing point where they could ford the river, all six men dismounted their horses and quietly led them across the stream so as not to attract the attention on the other side. Once on the other side, the Biloxi Kid gave the order to get on their horses and follow him to Sage Weed. Mounting their horses just on the other side of the river, they notice a large boulder. "Let's meet back here at this large boulder no later than 3:30 am. Have I made myself clear, soldiers?" he orders while pointing at it. "It is only about 3 miles to Sage Weed," he tells them as they ride off in a northerly direction. So far, they had been undetected, and once on the road heading to Sage Weed, if anyone saw them, they wouldn't think it unusual to see six men on horseback riding towards the town. When they were within a mile of the town limits, the Biloxi Kid brought the six-man patrol to a halt and gave them their last orders. "Once we get to Tuscaloosa Lucy's Saloon, I want us to split up when we get inside and start asking around if anyone had seen any Union Soldiers or if a man with a wooden leg and very noticeable limp had been there, and if so, how many men were with him.

"Captain, one soldier interrupted, can we fraternize with any of the pretty ladies we might run into inside the saloon?"

The Biloxi Kid caught onto the facetious attempt at humor and replied, "If Y'all have the money, then go for it, just remember we have to leave Sage Weed by 0400 hours so we can get back across the river to our camp while it is still pitch dark out."

"Yes, Sir Captain D'Frain, 3:30 am, we'll be here," the soldiers responded.

Johnny Spencer was leaving the tent of the commanding officer of the Union Army, Major Howard Stratton, which was encamped on the Texas side of the Rio Grande River. Johnny had been briefed that they were preparing for an all-out attack on the Confederate Army forces commanded by Colonel Bartson, scheduled to start after midnight on May 12, 1865. "Remember, Colonel, if you capture the Biloxi Kid, don't harm him, turn him over to me. I want to repay an old debt. "In fact, I will give you the $300.00 bounty. I would have been paid if I had captured him. In his case, money doesn't matter to me. "I want him, he's mine," he said as he started riding back towards Sage Weed.

Lucy spotted the Biloxi kid as he walked through the swinging doors of the saloon. Rushing over to him as he sat down at the bar, she called for him. "Hey, kid, Biloxi, Captain Jeb. I need to talk to you right away."

"Good evening, Lucy, good to see you."

"Good to see you too, Jeb," she says, grabbing his hand.

"What's on your mind, Lucy? You seem stressed."

"Yesterday afternoon, a group of about 20 men came into the saloon. "I started a conversation with one of them, a mean-looking man, and he had a wooden leg. "The more he drank, the more he blabbered, and then he pulled out a wanted poster that had your picture on it and asked if I had seen you. "He was carrying a powerful grudge against you, mentioning something about Gettysburg and you being responsible for him losing his leg.

Smiling, the Biloxi kid replies, "That would be Johnny Spencer, also known as Johnny One Leg."

"Was he right? Did you cause him to lose his leg?" Lucy asks?

"No, not directly. He was a soldier fighting for the Union Army at Gettysburg, and he lost his leg when one of the cannonballs fired in his direction exploded near him and blew his leg off. "Sadly, he was one of thousands of men on both sides who were killed or wounded in that battle."

"Well, kid, he seems to have shifted the entire blame on you, and he wants me to tell him if I see you, he says he has a personal score to settle with you. "I lied and told him I had never seen you."

"Thank you, Lucy. Eventually, I will have to confront him and settle, as he says, the score he feels he must settle."

"Be careful, kid. I don't think he will be happy until he kills you. "He rode out of here this afternoon, and I am sure he will be back again as the rest of his men are still in Sage Weed."

"That I'm certain of Lucy. He wants to be the one to kill me. "Thanks for the information. It's past one in the morning, and I must get back and meet with my men until the next time we meet, Lucy," he says as he takes her right hand and kisses it.

"That proves you are indeed a southern gentleman, Biloxi Kid, along with the Cajun French blood coursing through your veins. "I wonder what other French traits have been instilled in you, especially in the boudoir," Lucy asks as she presses her buxom body against him and teasingly kisses him on the lips.

The Biloxi Kid arrives back at the boulder where he is to meet his men and waits behind it. Soon, he hears the splashing sound of someone crossing the river on horseback. "Quiet, old gal, we don't want whoever that is to know we are here." It was still very dark and impossible to see the horse and rider as they approached the boulder, but he could hear him cursing that he had gotten his wooden leg wet and needed to dry it as soon as possible so it wouldn't become water-logged. "Damn, it's Johnny Spencer he muttered as the rider rode swiftly by and didn't notice him. It was now after three in the morning, and soon the other members of his patrol would be back at their designated rendezvous point. One by one, the other five men from the patrol arrived at the large boulder meeting point. All of them left the site promptly at three thirty AM heading back to the Mexican side of the Rio Grande River where Colonel Barston and his regiment were camped, and promptly at 0900 that morning the Biloxi kid gave Colonel

Barston the information that the Union Army was planning to attack the Confederate Forces at Palmito Ranch in Texas, which as history would later record as being the last battle fought in the American Civil War.

Major Stratton and his Federal Army force started their attack on Colonel Barston's Confederate forces after 8 am on the morning of May 12, 1865. Soon, Johnny Spencer and his 20 bounty hunters had combined with Major Stratton's soldiers as they pushed towards Palmito Ranch, running into scattered resistance from the rebels. They were successful in driving a wedge between the major force commanded by Colonel Barston and the battalion of soldiers that had the Biloxi Kid as their officer in charge. Johnny Spencer's Indian guide recognized him and rode his horse over to Johnny's position. Telling him he may soon settle his score with him. "Good news injun, I guess you're not as dumb as I thought you wuz," he tells the guide.

This bothered the Indian guide even in the heat of battle, "You wrong Johnny Spencer, I am not wise to the white man's ways, but I am not dumb," he angrily replies, and was close to calling him Johnny One Leg again.

Glaring at the Indian, he angrily says, "injun, I don't have the time to argue with you right now, but I promise you after I kill the Biloxi kid, I'll kill you next, you can bet all the wampum you have or will ever have. "Now get your red behind out of here, or I won't wait until I kill the Biloxi kid. I'll kill you now," he yells.

The Indian guide hits his pinto on the rear and speeds off, away from Johnny Spencer. "Me no think so, Johnny One Leg," he utters to himself, "me no think so," as he hears him laugh.

"You just wait, you dumb injun, you'll see. Nobody messes with Johnny Spencer and gets away with it," he says, laughing even louder.

As the day progressed, Major Stratton's Union Army forces and Johnny Spencer's bounty hunters were confronted by a large Confederate force led by the Biloxi Kid's infantrymen that had been reinforced by artillery and were forced to retreat from Palmito Ranch. Realizing the Biloxi Kid was leading the counterattack, Johnny Spencer took ten of his best men and rode in the opposite direction from Major Stratton's troops. He and the 10 men would lie in wait for the Biloxi kid near the Rio Grande River, where they would cross back into Mexico. "Soon will be eyeball to eyeball, and you're mine, Biloxi Kid. "Let's see how terrified you will be and how loud you can scream as I cut off your left leg and then the right one as an added measure of revenge just before I plunge my hunting knife deep into your Confederate chest. Then, once again laughing loudly, he says, "and I'll still get my $300.00 bounty money and won't have to share it with anyone. "I am going to be paid for doing something that I have sworn to do ever since that day back at Gettysburg."

Waiting in ambush on the Texas side of the Rio Grande River, Johnny Spencer and his men were alerted by the Indian guide who had ridden along with the men, "The Biloxi Kid and about a company size of Confederate soldiers are headed this way."

Johnny Spencer was unaware that the guide was even near, let alone riding with them. "Damn, I'm one hombre," he yells. "That just goes to show how dumb that injun is. I'm going to get to kill him and the Biloxi kid all at the same time."

Minutes later, the Biloxi Kid and his Confederate troops rode right into the line of crossfire that Johnny Spencer and his men had set up in ambush and were cut down like a fish in a barrel. The Biloxi kid dismounted his trusty stallion and slapped her on the rump. "Go old gal, go, there's no need for both of us to die." Then everything went blank as he was hit on the head from behind and fell to the ground. He didn't know how long he had been unconscious. He only felt like his head had exploded, and it, as well as his face and shirt, was covered with blood when he was brought back into consciousness by a bucket of water being thrown on his face. Coming to, he realized his hands had been tied behind his back and that his legs were tied together.

Sitting on a rock next to him was Johnny Spencer. "Well, well, Biloxi Kid, we meet at last, and now I am going to collect on a debt owed by you to me," he smilingly said as he gingerly ran a finger along the razor-sharp edge of his hunting knife. "Bet you wonder how I keep this ole knife so sharp, don't you? "I do it like this," he says as he rubs it back and forth across the grain of his wooden leg. "I guess it has some good use, after all. "But enough of this small talk. Now I'm going to cut off not one but both of your legs with it, and then I gonna shove it deep into your heart. "You got any last words, you rebel bastard," as he slowly walked over to where the bound Biloxi Kid was sitting up. Bending down, he cuts the trouser legs of the Biloxi Kid's pants and makes a surface cut on both of his legs. "It's payback time, Biloxi Kid," as he raises his arm and prepares to push the hunting knife into his left leg.

Feeling his life was about to end, the Biloxi Kid forces his eyes shut, and then, instead of feeling the excruciating pain of a 12-inch blade of the hunting knife being stabbed into his leg, he unexpectedly feels the body weight of Johnny Spencer as he falls on top of him. Opening his eyes, he sees his body, and with a tomahawk protruding from his back. Looking up and about 20 feet from where he and the dead Johnny Spencer were now lying, he sees the frowning Indian guide. "Not this time, Johnny One Leg, not this time. "Me not so dumb after all," he says as he walks and stands over Johnny Spencer and then unties the Biloxi Kid and sets him free.

The End

Christmas in The Lands of Our Ancestors

A Short Children's Story

by Ralph L Myers

Christmas in the lands of our Ancestors

Another Christmas Eve was here, and little Tess and Kevin's eyes were getting heavy with sleep; they could no longer be put off. Children in all the countries of the world were eagerly awaiting the arrival of Santa Claus. Putting down his book of children's Christmas stories after he had finished reading *The Night Before Christmas to his granddaughter Tess and grandson Kevin, Grandpa Ralph gently places his sleeping three-year-old grandson in his rocket ship-shaped bed, making sure the soft, fluffy blanket covered him*. Bending down, he gently kisses him on his forehead, making certain not to awaken him. "Sleep well, little guy, as you blast off in your rocket bed to some faraway dreamland. "Grandpa Ralph loves you very much," he whispers.

Then, walking back to the large, overstuffed chair where he had been reading stories to them, he carefully and quietly lifts his 5-year-old granddaughter and places her in her child's bed, with a fairy princess-shaped headboard and a dancing ballerina footboard. "Grandpa did our ancestors from Iran, Italy, Germany, and Greece celebrate Christmas like we do?" little Tess asks.

Christmas is observed in many countries, as it is here in America. "Tomorrow, after you and Kevin wake up and find out what Santa has left you under the Christmas tree, maybe we can explore how other countries celebrate this holiday. "In the meantime, close your eyes and sleep well, my beautiful princess. "Who knows, maybe the spirit of Christmas from all your ancestors' native lands will visit you and your brother in your dreams and tell you about the holiday as it has been celebrated in their lands. "I love you," he again whispers so as not to wake up her sleeping brother.

Yawning and rubbing her eyes, Tess says, "Grandpa, I hope that if the spirits of Christmas from our ancestors' native lands do pay us a visit in our dreams, they won't be scary like the spirits that visited Scrooge in the story you read to us yesterday."

"Don't worry about that, Tess, because you are only a child and you haven't been a bad person like Scrooge was. "Only good spirits will come and visit you in your dreams."

"Thank you, Grandpa, now I can go to sleep and not be afraid," Tess says as she closes her eyes and instantly falls asleep.

Shortly after midnight, as little Tess and Kevin soundly slept, each dreaming of Christmas day, a magic spirit placed itself in their subconscious so that both would be dreaming the same dream. "Don't be afraid, Tess and Kevin, I am the good spirit of Christmas, I am called Joy and Happiness. "I am here to take you on a journey in dreamland to the lands of your ancestors so you can see how Christmas has been celebrated in their native countries for thousands of years. "Take hold of my hand, and we will get started on our Christmas journey," Joy and Happiness tell them.

Good Spirit Joy and Happiness

Unafraid, Tess asks, "Where are we going, good spirits Joy and Happiness?"

"The first step in our journey will take us to Italy, where you will be given a tour of your Italian ancestors' native land and how they have celebrated the Christmas holiday. "There you will be joined by your cousins of many generations ago, Alessio and Alessia, and they will tell you about how they spent their Christmas in the past and how it is still celebrated in the present."

"But Mama and Papa will worry when they find we are gone, good spirit Joy and Happiness," Kevin says.

"Don't worry, little Kevin, if they look in your room, they will see that you and Tess are still here in your beds. "You will only be going on our magical journey in your dreams, and you will be back before you wake up," good spirit Joy and Happiness replied.

"Thank you, good spirit, we don't want to miss Christmas morning with our Mama, Papa, Grandpa, and Grandma," Tess says.

Flapping her wings together rapidly as a hummingbird does, sparkling fairy dust appears and covers little Tess and Kevin, and soon they are being whisked from the present to times long ago and the land of Italy, which is far, far away. Moments later, they can see little villages and happy people below as they fly far above them with the good spirits, joy, and Happiness. "We are here, Tess and Kevin, Italy, and soon I am going to turn you over to your cousins of many generations ago, Alessio and Alessia. "They will take you on your Christmas journey of Italy and tell you about its customs and traditions." Before Tess and Kevin could say anything, they were joined by a little boy and a girl. "Bongiorno, hello, I am Alessio, and this is my sister Alessia. "We are here to take you on a journey of our native land, Italy, and tell you how we have celebrated Christmas for many generations."

"Hello, I am Alessia, and welcome to Italy."

"Let's get started," Alessio says.

"Although it is now Christmas Day in Italy, Advent Sunday marks the beginning of the Christmas season in Italy. "Starting on this day, there will be many events and celebrations taking place, and then on nine days before Christmas, a time we call the novena, the Italian children will go house to house reciting Christmas verses in hopes of receiving coins," Alessia says.

Alessio and Alessia

"And on the first day of novena, many families will set up a presipio, or a manger scene, and they will gather before it each morning or evening to light candles and pray," Alessio tells them.

"Did you know that Christmas carols and manger scenes originated in Italy?" Alessia remarks.

"During the nine days of novena, Italian boys and girls write letters to their Mama and Papa wishing them Merry Christmas. "In those letters, they promise their parents they will be good, and they also make a list of gifts they hope to receive. "The letters are read at dinner time by their Mama and Papa aloud, and then they are tossed in the fireplace as they chant to La Befana, who is the mythical Good Christmas Witch, as their wishes go up the chimney," Alessio tells them.

"I become even more excited on Christmas Eve when I see the first star in the evening sky," Alessia says.

"Why, what happens that is so special that makes you more excited when you see the first evening star in the sky?" Tess asks.

"In Italy, as in many countries in the world, it is a part of our custom, culture, and belief that on Christmas Day, a baby named Jesus was born. "Not all people in the world believe this story about the baby Jesus, and I am not trying to get you to believe our story. I merely want to tell you about our customs and traditions as they relate to what we believe. "When the first star appears in the evening sky on Christmas Eve, every family puts lighted candles in their windows to light the way for the baby Jesus. "Also, they light candles and put them around their presepio or manger scene and pass the figure of the baby Jesus from one person to another and then place it gently in the manger," Alessio proudly says.

Father Christmas

"Yes, and then we have a large feast of fish or seafood served with lots of vegetables, salads, antipasto, bread, pasta, and sweets. Then later at night, we all go through the streets of our village that are lit by torches to church for the Christmas Eve Mass," Alessia cheerfully proclaims.

"Does Santa come and bring you lots of presents?" Kevin asks.

"Some Italian children receive gifts on Christmas from Babbo Natale or Father Christmas, which might be like this Santa Claus you ask about, but most have to wait until January 6, when La Befana, the good Christmas Witch, will bring them candy and gifts. "Usually on Christmas day, we celebrate only with our family, have another feast, and go to church. "One thing that is certain is we love the food that is prepared, especially Pannetone, which is a cake, and Panforte, also another kind of cake made from honey and spiced with cinnamon and cloves," Alessio tells Kevin and Tess.

"From Christmas Day and for the next few days until New Year's Day, we celebrate the holiday, and family and friends get together and visit, and the adults exchange gifts with each other, but as Alessio has told you, we children have to wait until January 6 to get our gifts from La Befana," Alessia says.

"On the night before January 6, Italian children will set their shoes on the fireplace, hoping they will receive the gifts they had asked for in their letters they wrote to their parents during novena, as La Befena only leaves candy and gifts for children who are good," Alessio remarks.

As Alessio speaks, the good spirits Joy and Happiness reappear. "Now you know about Christmas and how it is celebrated in your ancestors' native land of Italy. "Now we will pay a visit to Greece and learn about the Christmas customs and traditions observed by your Greek ancestors," Joy and Happiness says as she whisks Tess and Kevin away from Italy on their way to Greece.

Little Tess and Kevin suddenly find themselves flying high above the ruins of the Acropolis and Parthenon in the ancient city of Athens. "Look below, children, we are over the country that is the birthplace of democracy as we know it. "Also, it is home to some of history's greatest thinkers, mathematicians, and philosophers, Aristotle, Plato, and Socrates," the good spirit Joy and Happiness tells them.

"Is it Christmas time here, good spirit Joy and Happiness?

"Yes, Kevin, it is, and soon your Greek ancestors, Agathias and Agatha, will take you on a Christmas journey like your Italian ancestors, Alessio and Alessia, did. "They will show you and talk about the Christmas customs and traditions in Greece," Joy and Happiness answers.

Moments later, Tess, Kevin, and the good spirits Joy and Happiness are joined by a little Greek boy and girl, "Yia sou (yah soo,) hello, I am Agathias, my sister Agatha, and I will tell you how Christmas is observed and celebrated in our native country, Greece."

Agatha and Agathias

"Tess, Kevin, I will return for you later after your Greek ancestors, Agathias and Agatha, take you around their country and explain to you about Christmas in Greece," Joy and Happiness tells them.

"Let's get started, there is much to learn and see about Christmas in Greece," Agathias says.

"By December 6, Christmas is usually well underway. "During the "Feast of St Nicholas", presents are exchanged. "The Christmas celebration will last through January 6, which is the last day of the Feast of the Epiphany," Agatha adds.

"I think our Italian ancestors observe the same December 6 through January 6th Christmas celebration, or at least something as you do in Greece," Tess comments.

"What about Santa Claus, does he bring presents to the little Greek boys and girls?" Kevin asks.

"I think St Nicholas is like the person you call Santa Claus; however, in Greece, he is the patron saint of sailors. "His appearance is different because, according to Greek tradition, his clothing is drenched with brine and his beard drips with seawater," Agathias replies.

"Please tell us more," Kevin asks.

"Let me tell them," Agatha pleads.

"Okay, Agatha, you tell them about the solemn traditions, and I will tell them about the mythical spirit Kallikantzari. "You can tell your story first," Agathias says.

Thank you, Agathias. Here in Greece, Christmas is a very solemn holiday that celebrates the birth of the Christ child. "Beautiful carols or kalandas, which have been handed down from the Byzantine times, are sung. "On Christmas Eve, village children go from house to house offering good wishes and singing kalanda or carols as you would call them," Agatha says.

"Agatha, don't forget the small metal triangles and little clay drums the children use when singing the kalanda's," Agathias reminds her.

"Agathias, please, I am telling Tess and Kevin this part of the story, let me finish," Agatha impatiently says.

"I'm sorry, Agatha, please continue your story," Agathias tells her.

"Again, thank you, Agathias. "As I was telling you, when the little children go from house to house and sing the kalanda, they are often rewarded with sweets and dried fruits," Agatha says.

"I haven't seen many Christmas trees while we have been travelling through Greece," Tess observes.

"In Greece, Christmas trees are uncommon. "In almost every home, the main symbol of the season is a shallow wooden bowl with a piece of wire suspended across the rim that has a sprig of basil hanging from it wrapped around a wooden cross. "A small amount of water is kept in the bowl to keep the basil fresh and alive. "Once a day, a family member, usually the mother, dips the cross and basil in some holy water and uses it to sprinkle water in each room of the house," Agatha says.

"Why does the mother do this?" Kevin asks.

Killantzaroi Goblin

"I'll answer that question, Agatha. "This ritual is done to keep the Killantzaroi, or goblins and spirits, away that only appear during the twelve days from Christmas to the Epiphany on January 6," Agathias tells Kevin in answer to his question.

"That sounds scary. Are there really goblins and spirits here in Greece at Christmas time?" Tess asks.

"Myth has it that the Killikantzari only prey upon people during the twelve days of Christmas, from Christmas Day through the Epiphany on January 6. "It is believed the Killikantzari are usually male, and some believe they wear wooden or iron boots to kick people. "Others believe they take the form of wolves or even monkeys. "One story tells of a wicked stepmother who forces her stepdaughter to walk alone to a mill through the twelve days of Christmas because she is hoping the Killikantzari will snatch her away," Agathias says.

"On January 1, which is St. Basil's Day, more gifts are exchanged. "Also, on this day, the "renewal of water" takes place, which is the ritual when all water jugs in the house are emptied and then refilled with new St. Basil water. "The ceremony is often accompanied by offerings to the naiads, or the spirits of springs and fountains," Agatha says.

"Now you know much about Christmas celebrations, traditions, and customs in your Greek ancestors' native homeland of Greece," Agatha says as the good spirit Hope and Joy returns to take Tess and Kevin to their next land of their ancestors, Germany.

Once again, flapping her wings together and producing more golden magical fairy dust that soon covered Tess and Kevin, she says, "We're off to visit little Dieter and Erika, your German ancestors of long ago in the magical Christmas country of Germany."

In the blink of an eye, Tess and Kevin soon found they had arrived in Germany many years ago. A moment later, blond-haired and blue-eyed Dieter greets them. "Guttentag, (Good day) my name is Dieter, and my sister and I will take you on a tour of our homeland, Germany, and show you the customs and traditions we observe and celebrate."

Dieter

"Yes, and we have celebrated these customs and traditions for hundreds of years," little brown-haired and brown-eyed Erika says.

"Tess, Kevin, I will be back for you after you have taken a magical visit through Germany," the good spirit Hope and Change tells them.

Erika

"Tell me, where has the good spirit taken you so far on your journey?" Dieter asks.

"So far, we have visited our ancestors in Italy and Greece," Tess replies.

"Then many of the traditions celebrated in those countries are similar in many ways to what we observe here in Germany," Erika comments.

"The Christmas season begins in Germany on Advent, or 24 days before Christmas Day. "During Advent, our Mama bakes Stollen and Platzchen, which are Christmas treats and cookies. "What I like the most are the gingerbread houses and the hand-carved wooden Nutcracker (Nussknacker) figures," Dieter excitedly says.

"Here in Germany, like in Italy, Advent is observed. "However, in Germany we keep track of the days of Advent on 'The Advent Calendar,' which was invented in Germany and designed to involve all the German children in the festivities leading up to Christmas," Erika says.

"That sounds like fun. Tell us more about the Advent Calendar," Kevin says.

"Erika, do you want to tell Tess and Kevin about the Advent Calendar, or do you want me to tell them?" Dieter asks.

"You can tell them, Dieter, and if you leave something out, I will tell them about what you forgot to mention," Erika says.

"Okay, Erika, that sounds like a good idea, and I will try not to forget anything. "Each of the 24 days of Advent on the calendar is divided into twenty-four small windows, one for each day. "They are normally made of cardboard, but in ancient German times, the families would mark the 24 days of Christmas preceding Christmas with a chalk line on the wall. "In the mid-19th century or the 1800's, the first handcrafted Advent calendars were produced, and the first printed calendar appeared in Munich in the early 1900's. "Each day, a window on the calendar is opened which reveals a Christmas scene."

"I think I have seen an Advent Calendar at one of my friends' homes, and the little boy and girl that lived there showed me the chocolates and candy that were revealed when each new window was opened," Tess says.

"Tell us about Santa Claus, Christmas trees, and decorations in Germany. Do you have them?" Kevin asks.

"Yes, we do, and they are very beautiful. Let me tell you about the Christmas trees and decorations first. "In addition to having Christmas trees, in Germany we also have the Advent Wreath or Der Adventskranz, which has 4 larger candles and 19 smaller ones. "Each day of Advent, our Mama or Papa would light one additional candle, which would help us count the days until Christmas. "While this tradition originated in Germany, many countries throughout the world have adopted it. "When you visited your Greek ancestors, perhaps they also had Advent calendars, as many Eastern Orthodox churches use them. "The tradition of a ring of light existed among the Germanic tribes many centuries before the celebration of Advent began," Erika says.

German children always know about and look forward to the Christmas markets or Weinachtsmarkte, as they are called in Germany, which open in nearly every German town when the Advent season opens. "The town squares are lit up and alive with activity. "The townspeople gather together, listen to brass band music, drink beer, hot mulled wine, or apple cider," Dieter tells Tess and Kevin.

"Now, we've learned about Advent, the Advent Wreath, what about Christmas trees and Santa Claus?" Kevin asks again.

"In Germany, the Christmas tree is called Der Tannenbaum, and traditionally, German families usually put it up and decorate it on Christmas Eve. "However, some German families will put up their Christmas tree earlier during the Advent season. "The first known Christmas tree was set up in Freiburg in 1419 by the town bakers who decorated it with fruits, nuts, and baked goods. "The German children were allowed to remove and eat the treats on New Year's Day. Christmas trees are normally taken down on January 6, which is Three Kings Day, and eat any remaining sweets and treats they might have missed on New Year's Day," Erika says.

"I remember our Italian ancestors celebrated January 6, or the day that the good Christmas witch La Befana visits them and leaves candy and gifts for the children who have been good. "I also remember our Greek ancestors celebrate January 6 or the Epiphany," Tess recalls.

"You asked about St. Nicholas or Santa Claus, Kevin. "There once really existed a St. Nicholas who lived during the 4th century in a country that is now known as Turkey. Many stories have been told about him, and many legends are associated with him. He became known as the protector of children and anonymously gives them gifts. Since that time, the life and deeds of St. Nicholas have been celebrated on December 6 or St. Nicholas Day," Dieter says.

"In addition to there being a St. Nicholas in Germany, there is also a Santa Claus, or as he is called in German, 'Der Weihnachtsmann,' which when translated into English literally means 'the Christmas man." Santa Claus is a direct descendant of St. Nicholas, and in English-speaking countries, his name is derived from the Dutch term Sinterklass. "Many of the characteristics of the modern-day Santa Claus are easily recognizable in both the St. Nicholas figure and the personality descended from old German folklore. "The Weinachtsmann, much like Santa Claus, is a jolly old man with a long white beard that wears a red fur suit. On Christmas Eve, he leaves gifts for well-behaved children and punishes those who have been bad. "Kevin, I hope I have answered your questions about Christmas trees and Santa Claus," Erika says.

Soon, the good spirits Joy and Happiness return for Tess and Kevin, "We have one more country to visit before you will wake up in your beds in your rooms at your home on Christmas morning, and that country is Iran, where your Iranian ancestors are from."

"Thank you, Dieter and Erika, for telling us how our German ancestors celebrated Christmas," Tess says.

"Yes, thank you very much, Dieter and Erika, I have learned much about Christmas in Germany," Kevin adds.

Flapping her wings together for the final time of the night as Tess and Kevin are taken to the magical and ancient country of Iran, the land of their Iranian ancestors the good spirit Joy and Happiness tells them, "this will be the end of our journey in your dreams so you will learn of Iran's Christmas customs and traditions," she says as another magical golden cloud of fairy dust covers them.

In less time than it takes to blink an eye, Tess and Kevin arrive in Iran. Rubbing the magical fairy dust from her eyes, Tess asks, "Are we in Iran now?"

"Yes, Tess, you and your little brother Kevin are in Iran now, and I am Kurshed, and this is my sister Kursheed. "We are your ancestors from the ancient country of Iran."

"The good spirit Joy and Happiness have told us you have been on a long, long journey from your home in America to the lands of your ancestors in Italy, Greece, and Germany, where you have learned how the Christmas holiday is celebrated in each of those countries," Kursheed says.

Kurshed & Kursheed

"Come, hop on my magic flying carpet, and we will begin our journey," Kurshed says.

"What will keep us from falling off the magic carpet as it flies through the air? I don't see any seat belts that will hold us on the carpet that will prevent us from falling off," Kevin observes.

"Please don't worry, my carpet is magic, and it will not allow you to fall off with its magical powers," Kurshed tells them.

As Kurshed, Kursheed, Tess, and Kevin begin their journey through ancient cities and countryside in Iran, Khursheed begins to tell them about the Christmas holiday celebration in that country. "Iran is much different in how they celebrate or observe Christmas than the other countries the good spirit has taken you to visit. In your modern world, you came from Iran, which is a Muslim country, and Muslim's don't celebrate Christmas. "The major holiday observed by all Iranians, Christian or Muslim, is the holiday called Nowruz, a secular holiday on the first day of spring. "In ancient times, hundreds of years before the birth of the baby Jesus, who was born in Bethlehem and the Christmas holiday began, a man by the name of Zoroaster founded a religion that became known as Zoroastrianism, which believes in one God called Ahura Mazda. While the exact date of Zoroastrianism's origin is unknown, many believe it may date back as far as 1200 years before the birth of Jesus. "It wasn't until 549 years before the birth of the baby Jesus that the Persians (now called Iranians), led by Cyrus the Great, conquered Western Iran and founded the first Persian Empire. "History shows that Cyrus the Great was tolerant of all religions."

"Also, before the time of Jesus, there were popular winter festivals around Christmas time in many cultures. Also, solstices and equinoxes were important to many cultures, and most of the major gods in the Ishtar and Mithra religions had their birthdays on December 25, so there was a major holiday celebrating them on that day. "It was not until 221 years after Jesus' death that December 25 was the earliest year that proclaimed he was born on that day. "No one knows for sure why this claim was made sense it's clearly not winter in the Bible when Jesus is born," Khursheed says.

Since that time, cultures around the world have celebrated Christmas, and in Iran, people who are of the Christian belief celebrate it, which is more commonly known as the "Little Feast." "For the first 25 days of

December, a great fast is observed during which no meat, eggs, milk, or cheese is eaten. "It is a time of peace and meditation; a time for attending services at the church. "Christmas Eve is the last day of the fast, and on Christmas Day, people attend Mass, and then the Christmas feast begins," Kurshed adds.

"And just like in your other ancestors' countries, the Iranian Christians decorate Christmas trees, exchange gifts. "Also, even though Iran is an Islamic Republic and Muslim's believe that Jesus was a prophet, not a deity, as Christians believe, out of respect for those who celebrate Christmas, Christmas trees appear in store windows," Khursheed says.

"What about Santa Claus?" Kevin asks Khursheed and Kurshed.

"The boys and girls in Iran generally have not heard of Santa Claus, so they don't exchange gifts at Christmas, but they do receive new clothes, which they proudly wear all during the happy Christmas week," Kurshed answers.

"Soon, the good spirit Joy and Happiness will be here to take us back to our home before it is time for us to wake up on Christmas morning. "Kevin and I have learned so much about the Christmas holiday and how it is celebrated in the lands of our ancestors," Tess says.

Little Tess and Kevin could not remember the good spirit returning to Iran and taking them back home when they awakened from the deep sleep, they had been in. "Kevin, get up, it's Christmas morning, and we need to go downstairs and see if Santa Claus has left us presents under the Christmas Tree," Tess gleefully says.

"Tess, I had a wonderful dream that you and I visited our ancestors from Iran, Italy, Germany, and Greece, and they told us about Christmas in their native lands," Kevin says, hurrying to get dressed to go downstairs.

"So did I, Kevin, so did I. "We must tell Mama, Papa, Grandma, and Grandpa about our amazing dream and the journey to the lands of our ancestors."

The End-Grandpa Ralph

The Drone Incident

A Short Story
by Ralph L Myers

The Drone Incident

Somewhere in a secret underground military complex, "The coordinates are set and ready to launch Drone Seek and Destroy, sir."

"Authorization to launch is a GO, specialist.

Lifting the guard from the launch switch and pressing the green launch button, the specialist acknowledges the officer in charge, "Drone Seek and Destroy has been launched, sir, and is on course for the target at the latitude and longitude as designated."

Looking at the digital clock reading on the launch control panel, the specialist calls out, "It's 18:40 hours Zulu time, sir. Drone Seek and Destroy should reach the target at 03:00 hours Zulu time."

"Specialist, have telemetry reports recorded every 15 minutes for any potential course adjustments needed for it to reach the target; even a minor change could bring disastrous results if it goes off-course and impacts a populated area."

"Yes, sir."

A gray panel van with two men inside, parked not far from the secret underground military complex, slowly pulls out onto the road. The man in the passenger's seat dials a number on his cell phone. Listening intently, he hears the fax number dialing his cell phone. Once connected, he begins typing numbers and letters that are unintelligible, then presses Enter. Smiling, he turns to the driver as the van gains highway speed. "Success, we now are in control of the navigational system of the drone just launched," and then begins to enter more numbers.

Back at the underground military command center, the specialist who had just launched the drone calls out to the officer in charge. "Sir, something is wrong. I have lost control of the drone," he says as he excitedly tries to input data into the guidance system. "Someone has infected the program with a virus, and it has overridden our system and is changing course."

"Can you determine the new course of the drone, specialist?"

"Yes, sir, we can still get directional reading, but what is puzzling is that two new courses have been programmed into the drone, one is latitude 40-44'58" North and longitude 073-58" 04" West. "The other course is 38 degrees 53.3" North latitude and 77 degrees 02' 09.9" West longitude."

Entering the two courses the specialist had just read off into one of the NSA's satellites, the officer in charge gasps, "My God, those are the coordinates for the United Nations Building and the White House! "The President is scheduled to attend a meeting of the UN Security Council in about three hours; we have to contact him immediately and stop him from going to the U.N., as that is the drone's primary target."

"Sir, even if we are successful in alerting the President about the imminent threat of a rogue drone attack on the United Nations building, and he is still at the White House, and its longitude and latitude have also been programmed into it as the secondary target."

"How long do we have, Specialist, before the drone reaches its target?"

"We only have two hours until we have an impact in New York and 90 minutes until it reaches the White House, Sir."

"Specialist, do everything you can to regain control of the drone. I will alert the Pentagon of the impending crisis and get the President to safety. "Also, have your team trick chief contact the air defense headquarters and have them ready to scramble a squadron of F-16's upon the orders of the Pentagon. "If we can't regain control of the drone, we will have to shoot it down before it reaches either target.

"Mr. President and the United Nations, soon you will regret your very existence," the hacker mutters as he verifies the drone's target coordinates, continuing it on its destructive path.

Back at the secret underground military complex, the Officer in Charge is speaking to the Secretary of Defense.

"Yes, sir, Mr. Secretary, you heard me right, our Drone Seek and Destroy has had its guidance system compromised and will impact either the White House or the United Nations building in less than an hour. "We are doing everything we can to regain control of it," the officer in charge says as he is in a direct video conference with the Secretary of Defense.

"Give me an update every five minutes or as soon as anything changes, Colonel. Right now, our top priority is to get the President to safety. "The Chairman of the Joint Chiefs of Staff has been briefed on the situation and has put a squadron of U.S. Air Force F-16's on full combat alert. "They are in the air as we speak, awaiting final orders when you can locate the rogue drone. I am depending on you and your launch crew to regain the upper hand in this critical matter."

"Yes, Sir, Mr. Secretary," the officer in charge says as he walks back over to the Drone Command console. "Specialist, have you had any success in regaining control over Seek and Destroy?"

"No, sir, we haven't, and just when we think we can input our own virus into the drone's guidance system that will override the one that it has been infected with, the person or persons place a new one in the system," the specialist says.

"We are rapidly running out of time; it looks like our only option will be to shoot it down before it reaches its target. "Can you locate where the drone is, specialist?"

"Sir, since this is one of ours, it is very difficult to distinguish it from other aircraft in the area because our systems recognize it as friendly because of its electronic signature. "What we can do, sir, is order all aircraft in the area to switch their radar to a different frequency and then jam the drone's radar. "Once we do that, the drone will be picked up on our radar as unfriendly aircraft, and it will be easily identified and located."

"I'll give the order for all aircraft flying east of the Mississippi to change to SHF Band frequency for their onboard navigational systems, and at 19:00 hours Zulu time, you proceed to jam the drone's guidance system."

"Yes, sir, I understand and will commence with the jamming exercise precisely at 19:00 hours. Keeping his eyes on the digital clock, he begins to count down the seconds as the clock reads 18:59.50. "Ten, 9, 8, 7, 6, 5, 4, 3, 2, 1, jamming process has been initiated, Colonel."

Glancing anxiously at his monitor, the Colonel yells, "My monitor is blank, I don't see any aircraft, drone or otherwise, Specialist!"

The Colonel, no sooner had he yelled, could not detect anything; when a beep sounded, a bright blip appeared on the monitor. "Colonel, we've found her. The rogue drone has just entered the New Jersey airspace and will reach Manhattan and the U.N. building in approximately 12 minutes."

"Great work, Specialist. The F-16s should have it on their radar as well."

"This is Eagle one, I have locked on Drone Seek and Destroy and am ready to fire. Do I have authorization to fire?"

"Can you determine where the debris will fall, Eagle One?"

"This is Eagle One. If I fire within the next 10 seconds, my missile will impact the drone over the East River, and the debris should fall into the river."

"Excellent Eagle One, you have authorization to fire."

"This is Eagle One. The missile hit the drone, destroying it. It looks like the Fourth of July fireworks have come early to the good folks of Brooklyn, Manhattan, and across the river in New Jersey."

Jubilation erupted deep underground in the secret bunker, the Pentagon, and at the White House as confirmation was received that Drone Seek and Destroy had been shot down over the East River. "The Commander-in-Chief sends his thanks and congratulations for a job well done and for the countless lives that have been saved, perhaps his very own life," the Secretary of Defense says via satellite communication.

The End

What if The Pilgrim's had Landed on Plymouth Rock in 2011

A Short Story
by Ralph L Myers

What if the Pilgrims had landed on Plymouth Rock in 2011?

Every year, Americans commemorate the Pilgrims' landing on Plymouth Rock. What do you think the Pilgrims would have encountered if they landed in 2011 instead of 1620? Taking into consideration life as we are now experiencing it in the 21st century, I give you Plymouth Rock Landing 2025.

Plymouth Rock, 2025.

Headline: Boston Globe: A BOAT LOAD OF ILLEGAL IMMIGRANTS ARRESTED AT PLYMOUTH ROCK.

Story: The Department of Homeland Security and the U.S. Customs and Border Patrol greeted and immediately arrested the occupants of the Mayflower as they set foot on the shores of Plymouth Rock. An anonymous source from within the DHS, who asked not to be identified, told this reporter we suspect this is some fanatical terrorist group because of the way they dressed in a very unusual clothing style, black in color, with the men wearing stove pipe top hats and silver belt buckles. Also, we were told that the women must wear black dresses that leave no part of their bodies exposed below the neck, and that their heads must always be covered by a small white hat. An interpreter had to be brought in from Philadelphia who spoke fluent Quaker, as no one could understand what was meant when they kept saying "thee and thou."

Headline: New York Times: SUSPECTED TERRORIST LEADER MILES STANDISH, ARRESTED AND INTERROGATED WITHOUT PRESENCE OF AN ATTORNEY.

Story: The leader of the Pilgrims that landed at Plymouth Rock, Miles Standish, was whisked away to some unknown location by agents of the Department of Homeland Security, where he was interrogated and allegedly tortured by being forced to read aloud the more than 2500 pages of recently passed legislation. According to our sources in the State Department, many members of Congress have called for an immediate investigation to determine whether England is trying to undermine our economy by sending its poor and unskilled workers to work at the minimum wage.

After the Pilgrims were interrogated, they were sent to an illegal alien reception center located on Martha's Vineyard, a detention center that has long been a source of controversy between pro and anti-illegal immigrant rights groups.

A lawsuit has been filed in a federal court by civil rights groups on behalf of Miles Standish and the Pilgrims, calling for their immediate release and the closing of the facility.

Talk Shows.

"Larry King" - CNN: Mrs. Miles Standish was interviewed by TV Talk Show Host Larry King. Mrs. Standish told him that the Pilgrims had fled a tyrannical government in England in search of religious freedom. She said they merely wanted to practice their religious beliefs without fear of harassment or intimidation. A caller to the show, identifying himself only as a member of the ACLU, said he was filing a class action lawsuit on behalf of that organization against the Pilgrims for wanting to express their religious beliefs. "There should not be a place in American Society today for the expression of religious or moral values," the caller said.

"Bill O'Reilly," Fox News: The talking points on the O'Reilly Factor challenged the mainstream media and the kool-aid drinking left wing loons and pinheads' support of the illegal immigrant Pilgrims. "What a bunch of "Turkey's" Mr. O'Reilly said as he "opined" about a proposed national holiday being established commemorating the illegal landing at Plymouth Rock. "Illegal acts and blatant disregard for our laws should not be rewarded in such a manner.

Editorials: The Washington Post: "The Undocumented immigrant Pilgrims should be able to receive driver's licenses." "It would be to the government's advantage in keeping track of the Pilgrims if they are allowed to be issued driver's licenses. This policy raised immediate concerns among privacy rights organizations.

The National Transportation Safety Department raised additional safety concerns regarding Mr. Standish's question, "What's a driver's license, what's a motor vehicle?"

It does not matter, as they will ride their horses and drive their carriages on the wrong side of our roads, as that is what they are accustomed to in their native England, causing massive gridlock and numerous accidents. Also, what about the Amish that are already here legally? They drive their horses and carriage vehicles on the right side of the road as is how traffic moves in America. With addition of the Pilgrims who are customed to driving on the left side, one can expect numerous crashes, and deaths of hundreds of horses.

Counter Editorial: None could be found as the mainstream media does not permit opposing opinions because they are considered stupid, politically incorrect, and insensitive by the so-called intellectually elite.

Jay Leno, "The Tonight Show": "What about those Pilgrims, they land on Plymouth Rock illegally, and now they have the audacity to demand 'a piece of the rock." And if that is not bad enough, they want a no-smoking section of tables at next year's Thanksgiving Feast."

Entertainment Tonight: Newly arrived Pilgrim's Miles Standish and his wife Priscilla are taking America by storm, even though Miles was recently seen at a Thanksgiving Day bash with the exotic Native American beauty Pocahantas. When asked where Mrs. Standish was, Miles was in a daze. Could that sweet-smelling weed he was smoking in a pipe that was given to him by one of Pocahontas's friends have been the reason?

National Enquirer: The ghosts of Christopher Columbus and his crew from the Santa Maria were haunting the Mayflower as it sailed from England to the New World.

WELCOME TO AMERICA, PILGRIMS, IT'S 2025

Lost in The Forbidden Woods

A Short Story

by Ralph L Myers.

Pen Name for this story

Grandpa Ralph

Lost in the Forbidden Woods
By Grandpa Ralph

Once upon a time, in a make-believe village, a little boy and girl lived with their Mama and Papa in a house at the edge of the forbidden woods.

The little boy and girl were told by their Mama and Papa never to go into the woods just outside their yard.

They were told that many little children had gone into the woods and disappeared, never to be seen or heard from again.

The little boy and little girl were good children and always did what their Mama and Papa told them.

One day, the little boy and girl were playing in their yard with the little boy's football. Being a very strong little boy, he threw the football high in the air, and the little girl could not catch it, and it sailed over her head, out of the yard, and rolled into the forbidden woods.

"Oh no, I must get my football," the little boy told his sister, the little girl.

"No, no, you can't go out of our yard and into the forbidden woods," the little girl told her brother, the little boy. "Remember, Mama and Papa said we are never allowed to go into the forbidden woods because you will get lost and the wild animals will get you. "I'm afraid the football is lost; I can't even see it."

"I know Mama and Papa said never go into the forbidden woods, but I can't just leave my football in it."

Climbing the fence in their yard, the little boy runs into the forbidden woods. "Wait for me," the little girl says, "I don't want you to go into the forbidden woods by yourself."

The little boy held the little girl's hand as they entered the forbidden woods. "It's dark in here, and soon it will be night and even darker," he told his sister as he looked up into a dead tree with scary-looking night birds in it.

By now, it was so dark that the little boy and little girl could not see where they were going, and, like the little boy's football, they, too, were lost in the forbidden woods.

Walking farther into the forbidden woods, they heard a roar, then a strange sound that made them tremble with fear. The strange sound got closer to them, and the little boy and little girl held each other's hands very tightly as they tried to see a place in the dark where they could hide, but they couldn't find a hiding place.

"I wish we had listened to Mama and Papa and never had come into the forbidden woods," the little girl says.

"Me too," the little boy says. "I don't care about my football anymore." "If we ever find our way out of these forbidden woods, I will never disobey Mama and Papa again."

At that moment, Abe Ape saw the little boy and girl, jumped from behind a tree, and roared, "Why are you in the forbidden forest?"

"Oh, please, Mr. Ape, don't hurt us. We were playing with my football in our yard, and I threw it over the fence, and it went into the forbidden woods. "We were just trying to find it, and we got lost," the little boy tells Abe Ape.

"Don't worry, little boy, I won't hurt you. "I'm lost too, and I need to find my way back to the zoo. "Why don't we help each other find our way out of the forbidden woods?"

"Thank you, Mr. Ape, the little girl says, "We need to keep going so we can get out of the forbidden woods."

By now, it was very, very dark in the forbidden woods as the little boy, little girl, and Abe Ape continued walking, looking for a way to get out. "What's that, up in the trees ahead of us?" the frightened little boy says. "All I can see are shining eyes looking down at us."

Moments later, the little boy, the little girl, and Abe Ape heard something on the ground that seemed like it had jumped from the trees. Soon, they could see a Jungle Cat walking towards them, softly growling as it came closer.

"Little brother, what are we going to do? The Jungle Cat is getting closer, and he will probably want to eat us," the little girl says.

"I am stronger and bigger; I will chase him away," Abe Ape says.

"Go away, Jungle Cat, leave us alone."

Growling some more, the Jungle Cat says, "Do not be afraid, Mr. Ape, I will not hurt you, the little boy or little girl. "You see, I am lost in these forbidden woods and am trying to get back to the zoo. "Can you help me?"

"We are lost too, Mr. Jungle Cat, and trying to find our way out of these forbidden woods," the little boy says.

"Yes, and we are scared," the little girl says.

And I, too, am trying to find my way back to the zoo. Why don't you come with us?" Abe Ape says.

"Thank you, Mr. Ape, together, all of us should be able to find our way out of these forbidden woods," the Jungle Cat answers.

Together, the little boy, little girl, Abe Ape, and the Jungle Cat started again on their journey as they tried to find their way out of the forbidden woods. "Soon we should be home, I hope," the little boy said.

"And soon we should be back at the zoo," Abe Ape said about him and the Jungle Cat.

As they walked deeper into the forbidden woods, they heard a loud and frightening screeching sound from high up on the treetops. "Little Brother, what is that scary sound?"

"I'm not sure, but we had better stay closely together," the little boy replies.

Huddled closely together, the little boy, little girl, Abe Ape, and the Jungle Cat continued walking, carefully watching the tree tops rising above them. The screeching had stopped, but instead they heard the whooshing sound of flapping wings coming closer. "I think I have heard that noise before," Abe Ape says.

"What is it?" Jungle Cat asks.

Before Abe Ape could reply, a large Bald Eagle landed on a low limb of the tree they were standing next to.

"I am Edgar Eagle, and I am the bird that watches over these forbidden woods. "I used to live in the zoo, but flew away, and I hurt my wings when I landed too hard in one of these trees. "Now I can only fly as high as the top of some of the smaller trees, but when my injured wing is healed, I will be able to fly high in the sky, far above the forbidden woods."

Perched on a limb in a tree next to Edgar Eagle, Ollie Owl said, "Using my night vision and wisdom, I can help you find your way out of the forbidden woods."

"Oh, please, Mr. Owl, we would be so grateful if you would help us. I miss my Mama and Papa so very, very much," the little girl pleads.

"So do I, Mr. Owl, and my new friends also need to get back to their home at the zoo."

"Very well," Ollie Owl said, "and when you get back to your homes, I want you to tell everyone you were helped by me, Ollie Owl, the wisest bird in the forest."

"We promise, Mr. Owl," everyone in the group says.

"I will screech your wisdom from the highest mountain top," Edgar Eagle says.

"I will pound my chest and roar your great wisdom so all the animals in the zoo will know of it," Abe Ape boasts.

"So will the rest of us," the others promise Ollie Owl.

As they are led by Abe Ape further into the forbidden woods because of his strength and ferocious roar, and by Jungle Cat with its super senses of sight, hearing, and smell, and with Ollie Owl flying above them using his night vision, they soon come to a wide river that they would have to cross.

"Oh dear, there is no bridge for us to use to cross the river, and its current is too fast and strong for us to try and swim across it. What are we going to do?" the little boy asks.

"Edgar Eagle and I will be able to cross it because we can fly above the river, so we will try to find a place where you can safely get across. Come on, Edgar Eagle, let's go," Ollie Owl says.

Flying as high as his injured wing would allow, Edgar Eagle and Ollie Owl flew away from the others in search of a safe place to cross the wide, rapidly flowing river. "They will be safe there until we get back," Edgar Eagle says to Ollie Owl.

Waiting by the river's edge, the little boy, little girl, Abe Ape, and Jungle Cat began to think that Edgar Eagle and Ollie Owl must have gotten lost and would never return, since it seemed like hours had passed since they flew away in search of a safe place to cross the river.

They were about to give up hope when Edgar Eagle and Ollie Owl returned.

"Woo," Ollie Owl screeched.

"Don't you mean WHO, Mr. Owl, as owls usually screech WHO, WHO," the little girl asks?

"Yes, little girl, that is correct. I normally screech WHO, WHO, but in this case, I mean WOO, as I am exhausted after flapping my wings in flight while searching for a place to cross the river," Ollie Owl answers.

"I'm afraid we have bad news," Edgar Eagle says, "We could not find a bridge or any safe place where you can cross the river."

"What are we going to do? How are we ever going to get back to our home and our new animal friends to the zoo?" she asked as she began to cry.

Ollie Owl, being the wisest of the animals that were there, said, "Don't cry, little girl, let me think for a moment, just let me think."

As Ollie Owl sat perched on a tree limb just below where Edgar Eagle had gone, he tapped his head in thought as the others sat below him, very discouraged, feeling that they would never get out of the forbidden woods. "I am very sad," the little boy said, "I don't think I will ever be able to go home again."

"Mama was not telling us a lie when she said there were many little children that got lost in these forbidden woods and were never seen or heard from again," the little girl cried as large tears ran from her eyes onto her cheeks.

Just as all of them were about to give up any hope they would ever get out of the forbidden woods, they heard strange noises coming from the water in the river they could not get across. "What, or who is that?" Ollie Owl screeches as he points with his wing to the river.

"I see two objects in the river," Edgar Eagle screeches to everyone.

Before anyone could ask what, the objects were Turt the Turtle and Allie the girl Alligator, who climbed onto the riverbank from the river.

"I think we might be the answer to your problem and can help you get across the river," Allie, the girl Alligator, says.

Even though Ollie Owl was incredibly wise, he still had not thought of a way the others that could not fly would be able to get across the river. "Mr. Turtle, Miss Alligator, how can you help us get across this river?" Ollie Owl asks.

"Quite simply, your friends that can't fly across the river can ride across on our backs," Turt the Turtle says.

"Yes, that is the answer, and it is so simple I guess I didn't think of it because I am so wise," Ollie Owl says.

"That is quite true, Jungle Cat says, sometimes the answer to our problem is so simple we can't see it."

"Sort of like we can't see the woods for the trees," the little boy replies.

"What does that mean? I don't get it," Abe Ape says.

"Never mind, Mr. Ape, I will explain it to you after we get out of the forbidden woods," Ollie Owl tells him.

"Climb aboard our backs," Allie, the girl Alligator, and Turt the Turtle say, "we need to cross the river."

Soon, the little boy, little girl, Abe Ape, and Jungle Cat were safely across the river, thanks to Turt the Turtle and Allie the girl Alligator, as Edgar Eagle and Ollie Owl waited for them in a tree on the other side after they flew across.

Once on the other side, Edgar Eagle was extremely excited. "My injured wing is much better, and now I can soar high above the forbidden woods and see where your home and the zoo are, and I don't think it can be very far."

The little boy and little girl were incredibly happy, "Thank you, thank you, Mr. Turtle and Miss Alligator, you have made it possible for us to get home to our Mama and Papa," the little girl says.

"Yes, and thank you, our new friends Abe Ape, Jungle Cat, Ollie Owl, and Edgar Eagle, you will be back into your homes at the zoo," the little boy tells them.

Flying high above the forbidden woods, Edgar Eagle can see the little boy's and little girl's home and the zoo. "Follow me, he yells, I can see the zoo from your home and mine."

Following Edgar Eagle, the little boy and little girl soon found their way home and were happily reunited with their Mama & Papa and promised they would never go into the forbidden woods again.

Abe Ape, Jungle Cat, and Edgar Eagle were welcomed back to the zoo by Zeb the Zookeeper, where they told their amazing story of being lost and then finding their way out of the forbidden woods with a lot of help from Ollie Owl, the wisest bird in the forbidden woods.

Now you know the story of Lost in the Forbidden Woods and how the little boy, little girl, and their animal friends learned an important lesson: sometimes in life, we are tempted to do things and go to places that are forbidden. In this story, Lost in the Forbidden Woods, we see that all ends well for the little boy, the little girl, and their animal friends. However, terrible things can happen to us in life. We could end up regretting for a long, long time because we did not listen to our Mama or Papa. Sometimes it is too late to correct mistakes, and we must bear the consequences for violating rules designed to protect us from harm.

The End
Grandpa Ralph

Love Letters to High Pockets

A Short Story

by Ralph L Myers

Love Letters to High Pockets

The recently hired nurse orderly knocked softly on the door of room 106. Not getting a response to her knock she opened the door, entered the room and walked across the carpeted floor to where ninety year old U.S. Army Colonel (Retired) John (High Pockets) Packard was seated in his wheel chair staring blankly at the drapes that were partially closed on the window blocking his view of the Ohio River as its current carried a barge downstream.

"High Pockets, it's time for your nighttime meds," the nurse orderly tells him, causing a small smile to show on his aged and wrinkled face.

Slowly picking up his cane that he had hooked over the side of the wheelchair, he points to a bedside table and tells the orderly, "Put the meds there."

Walking over to the table, she placed several pills and a small glass of water that was easily within his reach. "Colonel Packard, as he preferred to be addressed, "how did you get the nickname high pockets?"

Turning in his wheelchair and facing the orderly, "You're new, aren't you?"

"Yes, Colonel, I am new, but I have worked in other veteran retirement homes, and I was told of a hero of Bastogne living here who was loved and respected by his troops, whom he commanded, and they respectfully referred to as high pockets. "Why did they call you high pockets?" the orderly repeated.

"Help me up, young lady, and I think you will see how I got that nickname."

The nurse wheeled him across the room to a wall with a wooden railing for support, like the kind used in a dance studio by ballet dancers. "Is this close enough, Colonel?"

"Close enough, young lady," he replied, taking his cane and hooking the curved end of it over the railing, and then began to pull himself up to a standing position slowly.

The nurse watched as the older man rose slowly. She was amazed when the now standing Colonel towered over her, even taking into consideration his 90-year-old frame was now partially bent over due to his age and osteoporosis he was suffering. "My goodness, Colonel, how tall are you?"

Smiling down at her, he remarked, "In my younger days, before Mother Nature and Father Time took over the aging process, I stood nearly 6 feet 7 inches. "I was much taller than all the men in my command, as well as the commanding officers in my battalion. "Just before D-Day in 1944, I had the privilege to meet General Eisenhower, who was 5'10," and I towered over him by nearly a foot. "When he shook my hand, he looked up at me and smilingly said, ' My, my Colonel, the pockets on your combat fatigues are sure up high. Then turning to his orderly, he remarked, "I sure hope he can keep a lower profile when we hit the beach at Normandy, otherwise Colonel High Pockets will be an easy target for the Germans. "The nick name high pockets stuck with me for the rest of my military career and when I retired Ike was President and he personally sent me a letter of congratulations and said he remembered that June day in 1944 when he met a young and very tall Colonel Packard thanking me for my service adding a post script "to Colonel "High Pockets" Packard."

"What an amazing story, Colonel. You were given a nickname by the future President of the United States, and obviously, you kept a low enough profile from the German snipers, or else you wouldn't be standing here today.

Sitting back down in his wheelchair, he requests the nurse orderly to help him get back to his bed so he can get the night medications. "Don't know why I should be so tired. I haven't done much today, just the usual breakfast, a few games of cards, watched some daytime TV shows, had lunch, and then back up here to my room. "Why can I remember leading my battalion on 20-mile marches and not feeling as tired as I do right now?"

"You didn't mention having dinner, Colonel Packard. Did you eat this evening?"

Sitting on the edge of his bed and taking the pills that the nurse had placed on the bedside table, a puzzled expression appeared on his face. "You know, I thought I did go down to the mess hall at 18:00 hours, but I can't remember for sure. Drumming his fingers on the bedside table, he struggled to remember, "Let me see, just let me see what they served?"

Trying to help him, the nurse orderly said, "Tonight they served rigatoni with Italian sausage, salad, and some Chianti, and then Spumoni for dessert."

Suddenly remembering and smiling, "Oh yes, now I remember the chow was unusually good for an Army mess hall. "Remind me to compliment the mess sergeant nurse."

It was obvious that Colonel High Pockets was confusing the retirement center's cafeteria with a military mess hall. The nurse orderly didn't want to correct him and softly assured him that she would pass his compliments on to the cook.

Colonel High Pockets, with the assistance of the nurse orderly, gets into bed. "Do you want me to turn out the lights when I leave, Colonel?"

Turning on the reading lamp above his bed's headboard, he says, "Yes, please, nurse, I just want to read for a while before I go to sleep." Once the nurse had left his room, he opened the drawer of the bedside table, carefully took out a bundle of old, frayed envelopes containing letters addressed to him, and removed an old, faded blue ribbon that held them together. Smiling, he opens the first envelope and takes out its contents and begins to read it out loud, as his mind wanders back in time as old memories are displayed in it like a movie he was now seeing again for who knew how many times before, yet it seemed as though each time he read it there would be yet another revelation of something that he had not noticed the last time. It was dated August 23, 1945, and was postmarked Antwerp, Belgium.

"My Dear Colonel High Pockets:

I hope this letter finds you well and happy. It has been almost 8 months to the date since you and your American Army soldiers found my papa, mama, and me hiding in the old barn behind our farmhouse. You will be happy to know that our home has been fixed, as you said. My papa and the other farmers around here have been helping each other put their homes and farms back in order, and just last week, all the farmers near the town of Bastogne had their best harvest since before the war began.

If you ever return to this area, please promise me you will visit us.

Love,
Claudine

Gingerly folding the note, placing it very carefully into the worn and tattered envelope, and setting it on the bedside table, he smiles as he remembers the first time he saw the beautiful young Flemish girl with long auburn-colored hair and a somewhat frail body. It was Christmas Eve, 1944, as Colonel High Pockets John Packard and the men from Charlie Company of the 82nd Airborne were being driven out of Bastogne towards the surrounding farmlands by a vastly superior and combat-hardened German battalion of SS Troops. The weather was bitterly cold, made even worse by the blizzard that was being magnified by almost gale force winds as they blew across the open fields, which were being pushed towards a farmhouse and the barn that was about 100 yards behind it. Yelling orders to his NCOs, he hoped he could be heard as artillery shells exploded all around them, and the howling, whistling wind made it even more difficult for them to hear. "Sergeant, take the first and second platoon and head for the house and secure it. I'll take the third platoon and take cover in the old barn, and we'll make our stand from there."

"You got it, Colonel, first and second platoon follow me," he yells as he starts running towards the farmhouse, "and kill anyone that gets in your way!"

Raising his long arm over his lanky body, "third platoon, run like Hitler himself is breathing fire down your necks and is still angry that Jesse Owens beat his superhuman race in the 1936 Berlin Olympics. "We've got to make it to that old barn about 100 yards behind the house," he yells while starting to run. "Ike, I sure as heck hope there aren't any German Snipers aiming for Colonel High Pockets," he thought as he recalled that day he had met the 5 Star General.

Running across open farmland, the first and second platoons were systematically being picked off by the German M80 machine guns being fired from the woods at the edge of the farm fields. Their positions were concealed by the trees, but they could be seen when they fired, as flames shot from the muzzles.

"Colonel High Pockets, sir, the fourth platoon should be in position soon and able to knock out those M80's," a corporal yelled as he caught up with him.

Running in a crouched position, trying to minimize the size of a target, he was with clumps of dirt being kicked up all around him by the bullets from the M80's. Colonel High Pockets yells, "Corporal, it had better be soon or else there won't be much of Charlie Company left."

Reaching the old stone walls surrounding the farmhouse, the 1st Sergeant ordered two rifle squads from the first and second platoons to take up a position and lay down a deadly field of fire directed at the dwelling. The rifle squads were literally disintegrating the brick-and-mortar structure of the house, 30 caliber M1 Garand rifles, and the BARs (Browning Automatic Rifles) as thousands of rounds struck it. "Cease fire," the 1st Sergeant orders as there wasn't any return fire coming from the house. "If there's any krauts inside, they're either dead or hiding someplace. "Corporal, take the third and fourth squads and set up a perimeter around the house. First and second squads, follow me, we are going on a little exploration adventure inside the farmhouse!"

There was no need to break down the doors or windows of the house to gain entry, as none were left after the barrage of firepower had been inflicted on them by the two rifle squads. Once inside, all floors and the basement of the house were thoroughly searched by the two platoons. "All clear, the house is empty, the Corporal reports to the 1st Sergeant."

"Not surprised," the 1st Sergeant replies, "but someone was here recently," he says, pointing to a still burning fire in the fireplace. "They were smart, that they probably got out when they saw us running across their field."

Loud explosions and then small arms fire could be heard coming from the woods, and then there was silence as the German M80 machine guns stopped firing. "The fourth platoon accomplished their mission and knocked out those Kraut machine guns," Colonel John "High Pockets" said as he and his third platoon troops surrounded the old barn at the rear of the farmhouse.

A young 1st Lieutenant and two squads of infantrymen cautiously and slowly approached the doors to the barn. "It sounds pretty quiet in the barn, too quiet, stay alert," he cautions, "there might be some SS soldiers waiting inside for us to come in. He orders one of the squad leaders to take 10 men with him. "See if you can get close enough to look inside, just try and keep your head down and not become an easy target for Gerry to pick you off if he's inside."

The squad leader and ten of his squad members, one of them using an old side mirror taken from a destroyed jeep, looked inside the doors that were partially open as streams of cold winter sunlight shone through the bullet holes, casting little beams of light onto the floor. "Sort of like stage lights shining on a theater's stage," he thought. "No one seems to be inside, Lieutenant," he calls back.

"Alright, squad leader, take your men and go inside," the Lieutenant orders.

By this time, Colonel High Pockets and the rest of the third platoon had taken up a position closer to the barn when he heard one of the men who had entered the barn shout out, "We found someone, probably the farmer, and there are two women in hiding with him. "Raus, he yells, (get out) and keep your hands up high." Soon, an older man and woman walk out of the barn, a confused look on their faces, but with their hands held high above their heads. They were followed by a strikingly beautiful young woman, 18 or in her early twenties, with long auburn hair.

Walking over to the man and two women, Colonel High Pockets, mustering what little German he could speak, asks them, "Sprechen sie English?"

The older man and woman looked more confused and somewhat fearful when they saw how tall the Colonel was, and turned to the young woman, speaking to her in what sounded like French to him. The young woman nodded her head and then looked up at the Colonel, who smiled and said, “I am Claudine, and this is my papa and mama. Turning and pointing to the house that was badly damaged and on the verge of collapsing, she said, “That is our home, we live here as did my grandpapa and grandmamma as well as their papa and mama before them.”

“Sergeant, make sure you leave some K rations with these folks; it looks like it’s been a while since they had anything to eat,” Colonel High Pockets orders.

“Yes, sir, Colonel, we’ll give them enough to last for about two weeks, as that is all we can spare until the weather breaks and more ammo and supplies can be air-dropped on us.”

The old man and woman smiled and thanked Colonel High Pockets as the beautiful young lady named Claudine stood on her tiptoes, motioning for him to stoop over a little, and kissed him on the cheek. “My papa, mama, and I thank you so very much, American G.I, Mr. Colonel High Pockets,” she laughingly said.

“No thanks needed, uh, I think you said your name is Claudine, miss. If anything, I am sorry that we have destroyed your house.”

The sound of a cart being pushed through the corridor outside his room 106 brought the Colonel’s mind back to the present. “I lost track of her after that as we continued our advance towards Berlin. “I wonder whatever became of her and her parents,” he thought while opening yet another letter he had saved for more than 60 years.

Unlike the letter he had just read from Claudine, this one was from a man, Issac A’brim, who had been a Jewish prisoner at the Nazi Neuengamme subcamp located on the outskirts of Wobbelin, Germany, which was northeast of Frankfurt, on the road to Berlin. It was dated June 18, 1945, less than two months after it had been discovered by the advancing 82nd Airborne Division troops on May 2, 1945.

My Dear Colonel Packard:

It has been 47 days since you and your American soldiers stumbled upon the Neuengamme concentration camp, where I and thousands of others were being starved, tortured, and systematically murdered by our Nazi oppressors. Had you not found us, I am most positive I would just have been another emaciated body cast into one of the many mass graves. I owe my life to you and our American liberators and for that you will and your men will always be in my thoughts and prayers as I daily say The Birkhat Ha-Gomeyl blessing is said after surviving illness, childbirth, or danger (including a hazardous journey or captivity).. You have my eternal gratitude.

Signed, Issac A'brim

Wiping tears from his eyes, Colonel John (High Pockets) Packard struggles for words as he always did when recalling or speaking about that horrible discovery he and his soldiers made on that day in May 1945. "I hope God has brought you some peace and comfort all the remaining days of your life, my friend, if you, like me, have lived so many more years." Carefully, he places the letter back into its envelope and puts it on the bedside table on top of the letter from Claudine.

Picking up another stack of letters that was bound by a red ribbon and all being from Barbara Ergmann, whom he referred to affectionately as "Babsi," the one and only true love of his life. Thumbing through the stack, he locates the letter postmarked from Kassel, Germany, dated New Year's Day, 1948, and takes it out, then, like the others, delicately unfolds it after removing it from the envelope. Holding the envelope up to his nose and breathing in, he can still imagine the fragrant smell of the perfume she always wore.

My Dearest Colonel John:

The haunting and melodic sounds of the Strauss waltzes are still playing in my mind. Dancing and being with you at the New Year's Eve party at the Officer's Club as we held each other so very close and then kissing at midnight when 1947 turned into 1948 shall always be a pleasurable and indelible impression on my thoughts and on my heart.

Although there were hundreds of others at the dance when I think of the two of us being together and waltzing the hours away it seemed as though we were the only ones there on the dance floor. I shall always treasure that evening and being with you, Johnny, and hope that you share with me those feelings of happiness and love.

Signed, Your Babsi

Folding and gingerly putting the letter back into its envelope he thought, "Oh how well do I remember that evening "Babsi" as that was also the night that you said yes, yes you would marry me and become Mrs. Colonel John "High Pockets" Packard. Then asking her as though she was in the room sitting beside him, "do you remember the night we first met at the Hauptbahnhof (main train station) in Frankfurt? "The war had just ended, and I was having a beer with a group of G.I.s that, like me, had been transferred to Frankfurt as part of the occupying forces. It was almost time to leave and head back to the Gutleut Kaserne when we all noticed four beautiful young Frauleins getting off a train. I turned to my buddies there with me and said, "Now this is what we just fought our way across Europe for," pointing to the four young ladies. One of the girls was at least 6 feet tall, perhaps a little taller, with what appeared to be natural blond hair. Being 6'7" myself, I was immediately attracted to her and told the others, "That fraulein is mine; you guys can fight over the rest of the girls if you want, but she is definitely mine."

"I have to agree, Colonel, she's beautiful, and I wouldn't mind getting to know her myself, but alas, I am only a Major, and you outrank me," one of the guys seated with me said.

"You're right, Major, I outrank you, and I am pulling my rank on you, so back off," I laughingly told him as I got up and started walking towards the four girls, all the while focusing my attention on the beautiful young blond. "I hope she can speak at least a little English," I thought as I was now standing directly in front of her and the other three girls. Before I could say anything, she smiled at me and said in perfect English, "My, my, you sure are a very tall G.I" This made the other three girls giggle like schoolgirls.

At least I now knew that she could speak English as I reached out my hand to shake hers and to introduce myself, "I am Colonel John Packard, but my friends and fellow G.I.'s simply refer to me as Colonel High Pockets. "May I ask you what your name is, Miss?"

Smiling back, she replied, "My name is Barbara Ergmann, and like you, I have a nickname; my friends call me 'Babsi."

"I would very much like to be your friend, Miss Ergmann. May I call you Babsi also?"

"If you would like, and I can also call you Colonel High Pockets," she replied with a mischievous grin.

Once more, holding the letter close to his nostrils and deeply inhaling it, trying again to get any scent of the perfume she wore, he smiled and said, "That night in the train station was to be the first of many nights they would spend together for the next 52 years," as a tear began to run down his cheek. "Oh, Babsi, precious sweet wife, I have missed you so much these past ten years since you were taken away from me, and since we never had any children, I can only share you with myself in my memories. "Be patient, my love. I think soon we will be together again."

The morning sun shone brightly into Colonel John "High Pockets" Packard's room as the orderly came in with his morning medicines. "Rise and shine, Colonel High Pockets," the orderly calls out, "time for your meds."

Not getting any response, the orderly walks over to his bed to awaken him. As he approached the bed, the Colonel was lying in it. He noticed the letters that he so carefully protected were strewn about on the floor, all but one, that was the letter from his beloved Babsi, which he was still tightly clutching in his hand that, along with his arm, was draped over the side of his bed. It wouldn't take a doctor or nurse to see that Colonel John "High Pockets" Packard had left this world and was now with his Babsi, leaving behind all the love letters to High Pockets.

The End
Ralph L Myers

Operation First Lady

A Short Story

by Ralph L Myers

Operation First Lady

It was nearing three in the morning on a warm, humid August evening. The moon was not visible as dark clouds formed overhead, being pushed by winds that grew stronger over Georgetown. The streets were empty, and it was eerily quiet for even this time of the night, especially considering how close it was to Washington, DC, as the unusual silence was sporadically interrupted by the sound of street cleaners and an occasional barking of a dog off in the distance.

Looking up at the sky and the gathering dark clouds Paper Bag Jones as he was called by other homeless drifters that knew him but didn't know his real name, because he always drank his wine from a bottle that was concealed by a lunch bag sized paper sack, utters to himself, "looks like it's gonna rain soon, I better get over to the White Cross Mission before the storm hits. "I can put up with some preaching by the Rev rather than getting caught outside in a downpour," he tries to convince himself as he stumbles across "M" Street heading towards the mission about three blocks from where he now was.

Stopping to catch his balance and to take another drink from the bottle of wine concealed by the paper bag, he noticed an older vintage Cadillac parked in the lot outside the entrance to the EZ- ZZZ's motel as a man about six feet tall emerged from the room next door to the motel check-in area. The man seemed nervous and, in a hurry, as he was led from the door of the room by four very large men who surrounded him and kept glancing from right to left and front to rear as they escorted him to the waiting Cadillac. Helping the man into the rear seat of the car, two of them got inside, seating themselves in the front seat, while the other joined him in the back seat. The fourth man rapidly walked across the parking lot and got into a plain black Ford sedan. Paper Bag Jones noticed the license plate on the Ford read GOV1 as it sped away from its parking place. As it passed him, he heard the driver speaking on what had to be a police radio: "CEO ONE safely onboard."

As the vintage Cadillac sped past Paper Bag Jones, he thought, "If he was being kidnapped, he was certainly cooperating with his kidnappers. By now lightning was lighting up the Georgetown skies and a streak lit up the Cadillac as it passed by and to his amazement he could see the occupant that the other men had escorted to the car, "my God, it's the President, Lucien Devens," Paper Bag almost shouted, nearly dropping his bag with the bottle of wine in it, "I've been hitting this rot gut too much lately. Realizing who and what he had just seen had a sobering effect on Paper Bag Jones, and he began to pick up his pace as the Georgetown skies opened and began drenching him and the area. Just after he passed the EZ-ZZZ's motel, a Yellow Cab drove into the driveway, and a very attractive, well-dressed young lady who appeared to be no more than 25 years old walked out of the motel room that he had seen President Devens leave, got into the cab, and sped away into the early Georgetown morning. "Well, well, well, that certainly wasn't the First Lady, Claire Devens," Paper Bag sarcastically says.

It was after 4 am when the vintage Cadillac discreetly pulled into an alley behind a remote parking lot that was not visible from the street near the White House. Getting out of the Cadillac, one of the Secret Service Agents opens the back door of the vehicle so that President Devens could get out. "Mr. President, Agent Rybold will drive you back to the White House in his Ford," he instructed.

"Thank you, Agent Bay," President Devens says, "as always, where I was tonight and who I was with will remain a classified matter."

"Yes, Sir, Mr. President," Agent Bay responds, "a state secret."

Once inside the Whitehouse, President Devens goes directly to the Oval Office rather than to his and the First Lady's living quarters in the Whitehouse. Opening the door, he is surprised to see an obviously intoxicated First Lady, Claire Devens, seated in his chair behind his desk. "Out with one of your bimbos again, Lucien," she asks in a slurring voice.

"Of course, not Claire, I was in Georgetown at the Ukrainian Embassy meeting with Ambassador Reigenovsky and Secretary of State Smathers on a matter of mutual national security interests between our two countries."

"You were there all night long? "Sure, and I was with the Pope convincing him that Priests should be allowed to marry," she angrily retorts as she staggers towards the door of the Oval Office, slamming it behind her.

The thunder showers that had inundated Georgetown, Washington, D.C., and the surrounding areas of Maryland and Virginia a few hours earlier did little to relieve the stifling heat. As morning rush hour approached, humidity was nearing 90%, making the commute into the city even more difficult as overheated cars caused a tremendous traffic jam on the beltway. Seated in the back seat of his chauffeur-driven Lincoln Town Car, Florida Senator Amber Harvest picked up her secure phone and dialed the office of House Speaker D.I. Stout, who had been working with her and other Senators to work out an agreement on the recently passed HR 32, the Disadvantaged Children's Relief Bill, to be sent to President Devens for his signature. After several rings, a voice came on the other end, "House Speaker Stout's office."

"Good morning, this is Senator Harvest. Is the Speaker in?"

"Yes, Senator, he's in. I'll put you through immediately," came the reply.

After a moment's delay, the Speaker answers, "Good morning, Amber, nice to hear from you, what's on your mind?"

Skipping formality and protocol, Senator Harvest gets straight to the point, "D.I., have you seen this morning's Washington Post, and then, before he could answer, continues, " It looks like the President is going to veto HR32?"

"Yeah, I saw it. I bet that really pisses off the First Lady, as she has been a staunch advocate of this legislation."

"Correct D.I., not to mention some of our very powerful backers that helped get us re-elected. "We can't afford to let them down, not if we want to depend on their continued support."

"Do you have any ideas, Amber?"

"Not really, D.I. I was thinking maybe the House Appropriations Committee might think of something to add onto the Bill that will sweeten the pot and change the President's mind."

"Amber, I think we're beyond that point because any amendments, no matter what they are, would meet with stiff opposition. "I know the Vice President is also in favor of the legislation, even though he is not married and has no children, but he and the President only seem to work together in front of the cameras and press. "Let me give him a call and see if he has any ideas that will salvage the Bill, and I'll call you back by the end of the week. "In the meantime, if you come up with anything, let me know immediately."

"Will do D.I., will do," Senator Amber Harvest says as she says goodbye.

After hanging up his phone, House Speaker D.I. Stout goes out of his office and tells his receptionist to please call Vice President R.L. Bellinger's office. "Stacy, call the VP's office and set up a time when I can talk to him."

"On the phone or in person, Speaker Stout?" she replies.

Thinking for the moment, he answers, "See if he can meet me in person somewhere up on the Hill; if not, a phone conversation will suffice."

"I'll do it immediately, Speaker Stout," Stacy responds.

In the shadow of Georgetown University, the phone rings at 34th Street and Massachusetts Avenue, NW, the Vice President's home, and is answered by one of his staff members assigned to work in the residence. "Vice President Bellinger's residence," the staff member replies.

"This is Speaker of the House Stout's staffer; the Speaker would like to schedule a meeting or telephone conversation with Vice President Bellinger."

"Please tell Speaker Stout the Vice President is scheduled to return to Washington tomorrow and will be hosting a conference in support of a variety of the President's programs he wants Congress to enact."

"Excellent, Speaker Stout is also scheduled to be at that conference. Could you please tell the Vice President he would like to take just a few moments of his time on a very important matter?"

"I will give the Vice President this information, and I am certain he can squeeze in a few minutes with the Speaker," the staff member replies.

At that very moment, First Lady Claire Devens, seated in front of a vanity mirror in room 1206 of the now infamous Watergate Hotel, carefully adjusts a wig with long flowing red curls that covered her short black hair. Looking at her reflection in the mirror, she smiles as Vice

President Bellinger walks into the room, and his image is reflected in the mirror, standing behind her. "We wouldn't want to start another Watergate scandal, would we, R.L.?" she asks, looking into the mirror.

"Certainly not, Claire," he responds while bending over and kissing the top of her wig-covered head.

"We'll need to meet again discreetly and very soon, I hope, R.L."

"I agree, Claire, the sooner the better, and in another location so our intimate liaisons will not be easily discovered."

Frowning, she replies more to herself than the Vice President, "If my philandering, no good, S.O. B. of a husband can chase after every skirt he wants, then I can at least enjoy being with a good man like you. "Who knows, if the word got out that I was playing around, maybe it would ruin him, but I don't want to be responsible for destroying your political career, R.L."

"Thank you, Claire, for that. I am very grateful."

"How were you able to get away with this meeting, R.L.?" Claire asks.

"I advised my staff I was in some delicate and secret meetings, but would be back as planned by tomorrow for the conference and could only be reached through the Secret Service. "What about you?"

"I told my secretary I was going to visit a dear friend in Virginia and would be driven there by my security. "Once outside of the White House, we stopped in a desolate location, and I changed into what you see me wearing now and had Agent Henry, whom I trust implicitly, take me here to the Watergate Hotel."

"Does he suspect or have any idea who you are meeting?"

"No, R.L., not in the least, and even if he did know, he would take a bullet before he would tell anyone," Claire answers.

Seated at a table in the far end of a Starbucks Coffee Shop out of hearing distance of any other patrons, two men, both well dressed, were involved in an intense conversation. Pausing for a moment, one of them pulled out some photographs and handed them to the other man. "Take a look at these pictures, Jack, and tell me what you think."

Putting his latte down and taking the photos from the first man, he carefully examined each one, "damn Bill, where have I seen that vintage Cadillac before?"

"It belongs to one of the secret service agents assigned to the President, and I've heard rumors, albeit unconfirmed, that the President has been seen riding in it," Bill replies.

"What the hell does the President need to ride around in that car for, even though it is a classic Cadillac, when he has an entire fleet of government limos available for his use? Looking closer, he points to a reflection in the rearview mirror. "Do you know who that is, Bill?"

"These pictures were taken in Georgetown the other night by some anonymous photographer who said he was staying at the EZ- ZZZ's Motel in Georgetown. "He said he thought it was some homeless person known around the area as "Paper Bag Jones," his real name is unknown," Bill answers.

"Interesting, Bill. We need to get a hold of Mr. Paper Bag Jones and see if he can tell us anything he saw that night. President Devens has powerful, influential enemies on the Hill who would probably be willing to pay a lot of money for any information that might discredit him. "Let's find him."

"Will do, Jack, and I think a good place to start will be at the White Cross Mission, as that seems to be the place most of the homeless use as a place to congregate and get out of the bad weather. "I'll get back to you tomorrow and let you know what Mr. Paper Bag has to say."

Taking a final sip from his latte and putting the empty cup down on the table, and then standing, Jack says, "I really look forward to hearing what you find out from this Paper Bag character, Bill."

It was getting close to dinner time at the White Cross Mission as Bill drove into the parking lot. "Dinner time should be a good time to find Paper Bag Jones if he's still around," he thought. He noticed a large police presence and crime-scene tape blocking off an area to the side of the building, which was constructed of typical red brick. Pulling it into the first available parking spot he could find, which the police allowed him to park in. As he opened the door and got out of his car, he was immediately approached by an unkempt woman wearing a ragged plaid dress that was probably three sizes too large for her.

"Hey, mister, got some spare change for me? I need to get some food."

Looking at the disheveled, grungy woman who obviously lacked the desire or availability to bathe or exercise any kind of adequate hygienic practices, Bill responds, "I could probably spare a couple of bucks, but since you're here at the mission and it's dinner time, I'm sure you can get a nourishing meal inside. "Besides, I think any money I give you will go to buy more booze or drugs."

Glaring angrily, the homeless lady turns as she says, "Go to hell, mister, you don't know me or anything else about me, so get off your pious high horse of judging me."

Feeling shame and more compassionate, Bill tells her, "I'm sorry, lady, and yes, my observation of you was indeed a value judgment, and my offer still stands, no strings attached or further questions, here's $5.00," pulling it from his wallet and handing it to her.

"Thanks, mister, turns out I misjudged you, too," she says, taking the paper Lincoln from him.

"I'm looking for a man who goes by the name Paper Bag Jones, lady. Do you know him or where I can find him?"

Examining the $5.00 bill as though she suspected she had just been given counterfeit money, she points toward the police crime scene tape. "He's over there," she says quietly.

"Where, over there, all I see are policemen and crime scene tape? Then it hit him like one of the red bricks from the White Cross Mission building had fallen on him, as he saw several crime scene photographers bent over taking pictures of what had to be a body. "What, he gasps, "do you mean that body over there is Paper Bag Jones?"

"You got it, mister, that's, or maybe I should say it was Paper Bag Jones."

A feeling of disappointment instead of loss or compassion flooded over Bill as he realized he and his fellow detective, Jack, had probably just lost any hope they would ever have of trying to find out who was in the vintage Cadillac the night the photos were taken, which he had been shown. "Wow, lady, any idea what happened? "By the way, what's your name?"

"I don't know what happened other than some of the regular customers here at the Mission said Paper Bag's body had not been discovered for two or three days until old Joe over there stumbled on it as he was looking through the trash for some cigarette butts, and my name is Hilda. "Joe said he heard the police telling the coroner it appeared as though he had been shot."

Turning, Bill starts to head back to his car. "Thanks, Hilda, I don't see a need to remain here any longer."

Before he could reach his car, Hilda once again approached him, "What'd you say your name is, mister?"

"I'm Bill," he replied, "why are you asking?"

Reaching into a pocket of her three sizes too large dress and pulling a folded and soiled piece of paper from her pocket she says, "mister, uh, Bill, it's not often someone comes here to the Mission wanting to actually meet and talk to one of us and your reason you wanted to talk to Paper Bag is none of my business and frankly I don't give a damn, but in our world friends are few and hard to come by and for some reason Paper Bag liked me and we would usually hang out together. "Two or three nights ago, during a loud thunderstorm, probably not long before he was shot, he ran over to me, out of breath. "Hilda, you're not going to believe who I just saw leaving the EZ-ZZZ's MOTEL, she recalls Paper Bag Saying. "It was raining so hard, and I was trying to find someplace to be that would help prevent me from getting soaked more than I already was, so I wasn't in a particularly good mood and sarcastically answered, ' Let me guess, the President."

"Damn Hilda, you see him too," his bloodshot eyes widening.

"C'mon, Paper Bag, why are you BS'ing me? What the hell would the President of the United States be doing in a rundown area like this?

"No B.S., Hilda, I saw President Devens riding in a white classic Cadillac, and there were maybe three or four very big guys with him. "I'm a little afraid because I think one of the big guys saw me. "Here, Hilda, take this. I wrote down everything I saw, and I want you to give it to the cops if something happens to me, handing her the dirty, folded piece of paper.

"What is it, Paper Bag? I remember asking."

"I don't want to say anything other than what I've already told you because I don't want anything to happen to you if something happens to me, so the less you know, the better and safer it will be for you," Paper Bag told me.

Intrigued, Bill asks, "Hilda, did you ever read what he wrote on that piece of paper?"

"No, I never did, as I thought Paper Bag was really bull shitting me, or the rot-gut wine he always drank had finally killed what remaining brain cells he still had left. "I guess he wasn't bull shitting me, or else he wouldn't be lying dead over there by the building with a bunch of cops taking his picture."

"Hilda, can I take a look at the piece of paper Paper Bag Jones gave you?"

"Sorry, Bill, Paper Bag said to give it to the cops if anything happened to him."

Pulling out his wallet and taking five $100 dollar bills from it and holding them up in front of Hilda, Bill says, "Hilda, I represent some very wealthy and influential people who would be willing to pay a lot more than what you see here if the information I think is written on that paper by your friend Paper Bag can be corroborated. "Just consider this $500.00 as a deposit with a large balance owed to you."

"You think maybe $ 100's more," Bill?

"No, Hilda, I think thousands more, and I will deliver it to you personally, in cash."

Hilda's eyes lighted up as though a cash register was going off in her head as she took the $500 dollars from Bill's hand and gives him the soiled piece of paper as she thought, "sorry Paper Bag, you'd do the same if the opportunity like this ever presented itself, besides the cops won't do anything with it except probably destroy it in order to cover up for President Devens."

Taking the soiled and folded piece of paper from her, Bill gives her his business card and says, "We'll be in touch, Hilda, and I promise you I will make good on getting the cash to you. "If you think someone is going to harm you, call me immediately, and I'll come and pick you up."

Bill couldn't wait to get back to his office, the P.I. on the Hill Detective Agency and go over the note's contents with his partner Jack which read, "I go by the moniker Paper Bag Jones but my real name is Talbert Riley and even though I am homeless and an alcoholic I swear I was in full command of my mental state on the night of August 11 when I witnessed President Lucien Devens leave the EZ-ZZZ's MOTEL in Georgetown, accompanied by three or four men in a white Cadillac. "One of the men got into a grey Ford Sedan with the license plate number CEO 1 on it. Moments later, I saw a very attractive young woman, probably in her twenties, leave the room the President had come from and get into a taxi. "As the white Cadillac left, one of the men in the car with the President saw me, and I'm not sure they will just forget I saw them and the president. If something happens to me, whoever finds this note, I want the police to have it." It was signed in surprisingly legible, neatly written handwriting: Talbert Riley.

"Bill, this is beyond belief if it's true and can be corroborated."

"Sure is, Jack, and worth a lot of money to the right bidder, of which there are probably many, considering this is D.C., and President Devens has a lot of enemies that would love to use this kind of information against him, especially, I bet, the First Lady."

"Check it out, Bill, and then we will let the bidding begin."

"Right away, Jack, and by the way, I promised the person who gave it to me a substantial payment for selling it to me instead of giving it to the police."

"How much, Bill?"

"I didn't say how much other than in the thousands."

"Hell, Bill, we can probably sell this information to multiple sources, pay your person the thousands promised, and still make a bundle."

"Yeah, and be doing a public service at the same time, Bill said, laughing in a facetious manner."

Speaker of the House Stout was meeting with Senator Harvest, Vice President Bellinger, and a reporter from the Post newspaper about the pending legislation President Devens had promised to veto rather than allowing it to become law. "Mr. Vice President and Senator Harvest, I have just been provided with some information that may make the President change his mind about vetoing the Disadvantaged Children's Relief Bill."

"Mr. Speaker, the only thing I can think of that would cause the President to change his mind about vetoing this Bill would be if you had pictures of him looking at kiddy porn in the Oval Office," Vice President Bellinger says.

"That would be too easy to spin away, Mr. Vice President. Then taking out some photos and a copy of a handwritten note from a large mailing envelope, pushing them face down across the table Speaker Stout continues, what you are about to see Mr. Vice President is something far more serious and damaging than a picture of the President looking at kiddy porn, it's photographic proof of his marital infidelity and possible, no probable evidence of his being involved in a murder of a homeless man in Georgetown just a few nights ago."

"D.L.," Senator Harvest interrupts, "not only would this mean almost certain impeachment of President Devens, but also conviction by the Senate that would probably be unanimous because no Senator in his or her right mind would vote against finding him guilty, regardless of party loyalty or favors they might owe to him. "We need to leak this information to the public immediately."

"I agree, Senator Harvest, and that is why I have invited a reporter from the Post to this meeting so he could help us do just that," Speaker Stout remarks.

"Mr. Speaker, Senator Harvest, while it is generally known the President and I do not see eye to eye on many of the issues and problems confronting this nation, I do think the First Lady should be made aware of this before she reads about it in tomorrow's Post," Vice President Bellinger says.

"Point well taken, Mr. Vice President. We'll ask our friend here from the Post to wait 24 hours before this news becomes public. "Since you brought up this point, who do you think should tell the First Lady?"

"I will, Mr. Speaker, all I ask is that you promise not to release this career-destroying evidence until after I talk to her and perhaps not at all if she showed it to the President and his mind is changed about vetoing the Disadvantaged Children's Relief Bill at her insistence and threat of exposing him in this matter."

"Fair enough, Mr. Vice President, even though nothing would please me more than taking down this President, I don't want to put the country and the international community through such a scandal and the potential ramifications this may have if this is allowed to run its ugly political and criminal course," Speaker Stout promises.

Putting the photos and the handwritten note back into the mailing envelope, Vice President Bellinger says, "The First Lady will be attending a luncheon I am hosting tomorrow on the Mall, and I will go over this information with her then, when I can get her aside in private."

"Mr. Vice President, the First Lady can keep them, as I can assure you there have been many copies made in the event these are lost, in addition to those in the possession of our source that made them available to my office, and at great cost."

Don't worry, Mr. Speaker, I get your point, and the information you have just given me will not mysteriously disappear," Vice President Bellinger says.

The luncheon had just concluded, and only a few of the guests remained as Vice President Bellinger approached First Lady Claire Devens. Both he and the First Lady's secret service agents were close by as he reached out his hand to shake hers, acting as though he rarely saw her instead of currently having an affair with her he greets her, "thank you for attending today's luncheon Mrs. Devens, by your support and influence much of your husband's policies would probably fail passage in Congress. "Now, if you could get him to support the Disadvantaged Children's Relief Bill, it would add yet another success to his legacy after he leaves office."

"Thank you, Mr. Vice President. I will try to use whatever influence I have to get him to sign the legislation when it reaches his desk. Then thinking "the only way I could influence him these days is if I arranged a ménage a Tre with him and two of his sluts."

"Before you leave, could you spare me a few moments? I have something I want to share with you," Vice President Bellinger says.

"Certainly, go right ahead," she replies.

"The information I have to share is highly sensitive and should only be heard by you. Looking around the room where the luncheon had just been held, Vice President Bellinger points to a corner in the rear near the speaker's podium. "Let's go over there," he says.

First Lady Devens, followed by her secret service agent, Henry, and the Vice President, along with his secret service agents, walk over to the table. Sitting down, he asks their agents to please allow him and the First Lady some privacy, "Gentlemen, please allow the First Lady and me a few moments to talk in privacy.

"Yes, sir, Mr. Vice President, we'll wait right over there," one of his secret service agents says, pointing to a table probably 30 feet away from where they were now seated.

First Lady Devens, noticing a look of concern on the face of her agent, said, "It's okay, Agent Henry, please just wait over there with the others. Once the agents had left and were not within hearing distance, she says, "Good to see you, RL, as she referred to him whenever they met in private. "What do you need to tell me that is so urgent and confidential?"

Taking the mailing-sized envelope from his attaché case, he hands it to the First Lady, "Here, Claire, see for yourself."

Taking the contents from the envelope and looking at each photo, first slowly and then rapidly as her face grew red with anger, but no expressions of surprise or disbelief, she put them back into it and slid it across the table to the Vice President. "This only confirms what I already suspected and the rumors I've heard about my cheating S.O.B. of a husband, and I'm not surprised that he would even resort to murder to keep his sordid life private. "Why are you showing me these photos and the note, R.L.?"

"I have been approached by some very powerful people up here on the Hill who want this information leaked to the media unless they can be used to influence the President not to veto the Disadvantaged Children's Relief Bill. "They have given me 24 hours to talk this over with you to see if you can use them to assure the President's cooperation in this matter. "If you can, nothing further will come of this, and all the photos and the note from the murdered homeless man will mysteriously disappear."

"R.L., my gut reaction tells me to go ahead and let the information become public and destroy the bastard, but my scorned woman instinct tells me otherwise, so tell whomever those powerful people are, I will handle the situation personally and not allow this information go public. The President and I are going to Camp David this weekend, and I will reveal to him just how much of a bastard, degenerate, and murderer he is. "Reassure them that the President will not veto their legislation."

"Okay, Claire, you have my word, I'll relay your assurances to them," he says as he hands the envelope back to her, "and Claire, they told me if what you have gets lost, they have many more copies they can use."

"They are serious adversaries, aren't they, R.L?"

"Yes, Claire, more serious than one could imagine."

At last, a break in the high temperatures and humidity that had been prevailing in the Washington, Virginia, and Maryland area arrived as the President and First Lady were being driven to Camp David by their trusted and loyal Secret Service Agent Henry. As they neared Camp David in a remote part of the Maryland wooded countryside, Claire took out the mailing envelope that Vice President Bellinger had given her two days ago at the luncheon. "Lucien, I have something to show and discuss with you."

"Can't it wait, Claire? I'm working on a speech I want to give in Boston next week."

"No, Lucien, it can't. We need to discuss it now, while we are alone and can talk."

Sighing, President Lucien Devens says, "Okay, Claire, just what do you need to talk about that is so important that it can't wait until we arrive at Camp David?"

"Here, take a look at this, you arrogant bastard," she angrily says, throwing the envelope in his direction.

Opening the envelope and thumbing through it, the expression on the President's face changes from irritable to surprise and finally to alarm. "Where did you get this? Who else has seen it? What do they want?" he demands.

"That's not important, you are a lying, cheating, no good bastard, other than they want you to forget vetoing the Disadvantaged Children's Relief Bill."

"Are you in on this, Claire?"

"Lucien, we both know about your womanizing and many indiscretions since you became President and probably years before, but no, even considering them, I was not involved."

Taking the photos, note, and mailing envelope, he tears them into small pieces and places them in the ashtray in the rear seat of the limo they were riding in, then sets them on fire. "Laughing, he says, "Where's your proof now, Claire?"

"Lucien, Lucien, Lucien, neither I nor the people who gave me this information are as stupid as you think we are. "Don't think for a moment those were the only copies, I can promise you they weren't."

"Claire, I beg you, think for a moment how this will ruin me and how scandalous it will be for you. We have to get all the copies back, no matter how much it costs."

"Lucien, I don't really give a damn about what it does to you, your career or legacy, and as for me I'll survive, which is saying a lot more than it does about you", as she then takes Agent Henry's service revolver from the arm rest where he had placed it, levels it at her husband, pulling the trigger, killing him instantly and then turning the gun on herself pulls the trigger again wounding but not killing herself.

Agent Henry takes the pistol from her, calmly wipes off her fingerprints, and tells her, "First Lady Devens, it has been my pleasure to serve you for the past several years. "Now it's time to part ways," he says as he gets out of the limo he was driving and pulls to a stop on the side of the road.

By now, several other cars with Secret Service Agents, including the vintage Cadillac owned by one of them, had reached the President's limo. Getting out of their cars, the driver of the Cadillac yells at Agent Henry, then seeing the pistol in his hand, yells "gun," and he and the other agents bring down Agent Henry in a hail of gunfire. Running over to the President's limo, he sees the President, who has been mortally wounded, and a conscious but bleeding First Lady. "Agent Henry has just killed the President and wounded the First Lady. Get his doctor riding in another car over here immediately to tend to her wounds," he yells.

Several months had passed since that fateful day in the Maryland countryside. It was now spring in the Nation's Capital, the cherry blossoms were in full bloom, and it was a beautiful and warm spring day as now President R.L. Bellinger stood at the front of the Rose Garden looking upward at the red velvet carpet that extended nearly the entire length of it. Walking down the red carpet towards him was Claire Devens, who in just minutes would make history by becoming the new First Lady Claire Bellinger for a second time as they wed at the White House. Everyone in attendance stood as Speaker of the House Stout accompanied her down the aisle to give her away to the awaiting groom, President Bellinger, as a beaming Senator Amber Harvest, her maid of honor, looked on.

Seated among the guests were one of the children who had become the poster child for the Disadvantaged Children's Relief Bill campaign and a former homeless woman known only as Hilda, now dressed in the most fashionable clothing money could buy. After all, she had earned it.

"Lucien, you bastard, I told you I would be okay, why I am still the First Lady!" she thought as her smile widened, as she suppressed the urge to laugh!

The End

The Augur Sketch Artist

A Short Story
by Ralph L Myers

The Augur Sketch Artist

Robbery Homicide Detective Matt Lelone, along with the Police Department's sketch artist Will Darbin, stood at the hospital bedside of a middle-aged woman, Elka Felice, whose badly bruised and swollen face bore witness to the savage beating she had suffered.

"Hello, Miss, or is it Mrs. Felice? I am Detective Lelone, and this is Will Darbin, our police department's sketch artist. "Do you feel like you can answer a few questions about what happened and give us some information about your attacker's appearance?

Struggling to look at the detective and sketch artist through swollen eyelids that were dark purple and blackened, in addition to her forehead and face, she replies, "It's Miss and I'll do my best but I don't remember much that happened during the attack just that when I came too I was on a gurney being placed in a paramedic unit."

"That is all we can ask Miss Felice, and if you get too tired or need a nurse or something to come in, just let me know, and I'll stop asking you any more questions. "By the way, when I'm done asking you questions, I am going to have you try and describe your attacker as best you can to Mr. Darbin here so he can draw a composite of them to help us in catching this thug. "You don't have to be formal, you can just call him Will. In fact, down at the station, we refer to him as Willy the Sketcher."

"Hello, Miss Felice, I'm Will Darbin, but as Detective Lelone said, you can call me Will or Willy, whatever you prefer. Taking my sketch pad from the portfolio I had brought with me and a #2 lead pencil, "Do you remember anything about the person that did this to you? "Was it a man, woman, teenager, was the person white, black, Hispanic or Asian," I ask while writing her name and the case file number on the sheet? "I am going to try to develop a composite sketch of the perpetrator who did this to you."

"As I said, I don't remember much at all after he hit me several times."

"Okay, that's good, Miss Felice. At least we have established it was a man. "Are there any other features or characteristics about this person that come to mind?"

After several minutes, Miss Felice had provided enough information that allowed me to develop a very rough image of what the assailant might look like. Holding up the sketch so Miss Felice could see it, "does this person look like your attacker?"

Beginning to cry, "Will, I wish I could be sure. "He could look like that. "I want him caught so he doesn't do what he did to me to another innocent victim."

"Thank you Miss Felice please get some rest, you have been very helpful and with the information you have given me I think I have enough to compare my sketch with mug shots but if you remember anything else, anything else at all I am leaving my card and Detective Lelone's on the bedside table, just give either one of us a call."

Gathering up the drawing and placing it in my portfolio, I again thank Miss Felice and walk out of her hospital room and into the hallway of the Emergency Room, which was in high gear with people in need of urgent care for one reason or another. "Must be a full moon tonight" I thought while walking out of the ER and into the parking lot to where I had parked my car. Starting the engine and pulling out of the hospital parking lot onto the street my mind wandered as I thought about poor Miss Felice, the hectic pace of the doctors and nurses in the ER, getting back to the station where I could compare the very vague and rough sketch of her attacker with mug shot pictures. I was so preoccupied in fact that I never saw the SUV that was careening across the road I was on after being hit by a large truck directly into my path. It was too late to react, there wasn't any way I could avoid hitting the out of control SUV.

A light shone brightly above me, and all I could see were blurred figures and the sound of voices coming from them. The pain I was now experiencing was like no other I had ever felt. "Surely, I couldn't be dead because I was always led to believe that once one dies, all pain and suffering stops," I reasoned.

"Can you tell me your name?" a masculine-sounding voice asks.

"Where am I? What happened?"

Ignoring my question, the masculine voice again asks, "Can you tell me your name and address, sir?"

"Will - - - Will Darbin, I live at 2395 8th Street, Apt D, here in Ashcamp. "What happened? Why do I hurt all over? Where am I?"

"Welcome back, Will. I'm Doctor Stillwell. You have been in a medically induced coma for three weeks. "You are in the Intensive Care Unit at Our Lady of Mercy Hospital, and we didn't think you were going to make it. "What's the last thing you can remember?"

The pain in my head seemed to intensify as I tried to remember something, anything. "You say I'm at Our Lady of Mercy Hospital? Unless I am totally confused and disoriented, it seems to me I was at this hospital visiting a crime victim, putting together a composite drawing of the person who attacked her. "I remember leaving her room and ER and getting into my car, paying the parking lot attendant, and then leaving the lot, and I can't remember anything else after that."

"Well, there's a simple explanation for that. "After you left the parking lot, your car was hit head-on by a large SUV that was pushed into your path by a truck that had hit it. "It was fortunate for you that you were literally just outside of the hospital's ER and received lifesaving immediate medical assistance because you would not have survived the accident otherwise. "Sadly, the woman driving the SUV was not as fortunate as you and died from her massive injuries at the scene."

"Was the accident caused by me?"

"From the report, the paramedics and police stated that the SUV driver's vehicle crossed the street into your lane and hit you after she was rear-ended by a large truck.

Now my head hurt even worse as I thought," I was here at the hospital trying to help a victim and as a result after I left the hospital I caused another woman to become a victim- -only she did not survive and even though it was ruled as being not my fault I don't know if I'll ever be able to block it from my mind that I was at least partially responsible."

It had been six months to the day since the accident, six months of excruciatingly painful rehabilitation and six months of psychological counseling, when I was finally released from Our Lady of Mercy Hospital's rehab unit. It felt good to be out and able to begin a new phase of life. Physically, I was probably about 90% of what I was before the accident. Mentally, I wasn't so sure, as no matter what, my psychological counselors through therapy had tried to have me understand that I had no control over what had happened to the female driver who was killed. I couldn't totally accept it, and I couldn't shake the ever-present sense or feeling of guilt. Now, and for the first time in my life, I was experiencing severe and at times almost debilitating headaches. "I imagine in time they will subside and probably go away. In the meantime, I will somehow have to deal with them," I surmised.

"Hey, looks who's finally back, Willy the Sketcher," Robbery Homicide Detective Matt Lelone calls out as I walk through the door of the detective bureau's office. "Glad to have you back, Will, for a while there it seemed like you might not survive, let alone ever be able to be back to work so soon."

Waiving to everyone, "It's great to be back, I just hope I still have a desk around here somewhere, Matt."

"Relax, Will, would the Vatican have thrown out and replaced Michael Angelo's masterpiece painted on the ceiling of the Sistine Chapel with a rendition of Dilbert? "I don't think so, and even though there have been other sketch artists assigned to the department in your absence, none of them could hold a candle to you, the Ashcamp PD's very own Michael Angelo."

Sitting down at my desk, still as cluttered and in a state of disarray as it was the last time I was here, I begin to thumb through some of the old files. Soon, I ran across a file that seemed familiar but was still vague. It was marked Ashcamp PD File 416754, victim Felice. "Help me out here, Matt. Wasn't this the victim I was interviewing with you the night of my accident?"

"Good to see your memory has returned, Will. That is correct; she is the one who had the holy crap beat out of her."

"Whatever happened to her? Did they catch the guy who did it? "Did my sketch help?"

"That was one of the most frustrating and sad cases I have ever been involved with. "In answer to your question, no, your sketch didn't help, but that wasn't because it was a poor one, as it turned out, it was bogus."

"Bogus, what do you mean, Matt?" I ask.

"A couple of days after the night we interviewed her, we found out that while she was a victim of a crime, the person who beat her up so badly turned out to be her significant other, her husband. "We found out about this by accident when one of the nurses overheard her talking to a visitor, a male, and assuring him she wouldn't tell the police that he had assaulted her."

"Why in God's name would she tell him that? "One of the things I now remember is her telling us that she wanted the person who attacked her to be caught so he wouldn't hurt another person."

"If I could answer your question, I would be some rich and famous TV Psychologist, an Ashcamp PD's version of Dr. Phil. "I can tell you this, however, in the 20-plus years I have been on the force, it is not unusual for a battered spouse to change her mind and not press charges. "In this case, the Felice case, she wouldn't press charges."

"I wonder how she is doing now, Matt."

Walking over to his desk, he takes a file from a stack and brings it back, handing it to me. "Here, see for yourself, Will, I didn't want to show you this, especially on your first day back to work, but since you asked, I guess now is as good a time as any."

Taking the file from Matt, the first thing I noticed was the words HOMICIDE INVESTIGATION with the name of FELICE as the victim, screaming out at me. Opening the file jacket, it revealed gruesome and horrible pictures of a badly bruised and mutilated female body. "My God, Matt, she looks worse, much worse than that night we interviewed her in the hospital. "Did her husband do this to her?"

"Yeah, Will he did, and he's in custody right now, and his attorney is trying to cop a plea bargain from the DA's office for him. "If only Miss Felice had pressed charges earlier, she probably would still be alive."

I could feel one of my migraine headaches coming on as I thought back to the many counseling sessions I had while trying to work through the guilt and issues that remained after finding out the woman who crashed her SUV into my car had been killed. "Believe me, Matt, I know how you feel. "My shrink kept telling me the woman that died in the crash I had her death was not my fault. "But somehow I can't shake the feeling that, at least in part, it was my fault. "I bet he would be telling you the same thing, and while he may be correct, nothing he can say will ever bring them back to life."

My first day back to work seemed like it would never end, but thankfully, it did. Fortunately, we were not called out to any new cases, and no one walked into the office to report being a crime victim. Walking through the door of my apartment, I put my keys on the end table at the side of the sofa and glanced at the phone to see if I had any new messages, and I felt relieved that the message light wasn't blinking. By this time, my headache was reaching an intense point, but now I felt a new sensation, a feeling I had never experienced or felt before. It seemed as though I were being drawn by some invisible force to my desk and the sketch pad I leave there to make sketches of suspects when the victim gives it to me over the phone. After I make one of these sketches, I scan it and then email the file to the person who described it to me.

Sitting down at the desk, I instinctively reached for the pad and took a #2 black pencil out of a tall black mug I had put there to hold pens, pencils, gum erasers, and other drawing supplies. Without thinking further, and no one prompting me about what a person they were trying to describe looked like, the pencil and my hand seemed to be anatomically connected, and I began to construct an image on the pad rapidly. I had no idea who the person was or why I felt this unavoidable urge to draw, or how I should or would know what they looked like. No matter what I did or tried, I had lost control of my hand and what was being drawn on the sheet of sketchpad paper. Finally, I regained control of my right hand and set down the pencil. Picking up the sheet of paper, the face of a young Caucasian woman with a fair complexion and light colored hair that appeared to be in her twenties stared back at me. "Who was she, and why was I somehow forced to draw an image of her?" Picking the pencil up again and writing the date and time I had drawn her, I started to put the pencil back into the black mug when the same powerful, strange feeling returned, and once again I was involuntarily drawing yet another figure on a different sheet of the sketch pad. When the new image was completed, and I was able to put the pencil down, the image of a small boy that was probably six or seven stared back at me. The image I had drawn was so detailed that it revealed he was missing his two upper front teeth. "What's happening to me? Am I losing my mind, am I dreaming, having a bad nightmare? Looking at and closely examining my right hand, I muttered, "Are you finished with me now? "Will, you need to call the shrink first thing tomorrow, something is drastically wrong and you must be going crazy." In a near state of panic, I retrieved a bottle of Jack Daniel's Tennessee Sour Mash Bourbon from a bottom drawer of my desk, opened it, and gulped down what easily would have been three or four shots had I poured it into a shot glass and measured the amount. I sat there waiting for the bourbon to take effect and when it finally did I walked precariously into my bedroom and collapsed on the bed not bothering to undress not even taking my shoes off and soon fell soundly asleep or perhaps a better term is passed out.

The loud and incessant ringing of my telephone awakened me from my Jack Daniel 's-inflicted stupor. It was 5:30 am and still dark outside. Managing to pick up the phone on my third attempt, I could answer the familiar voice of Detective Matt Lelone, who came booming through the earpiece. "Will, meet me at the ER of Our Lady of Mercy as soon as you can. "The paramedics just brought in three victims of a hit-and-run accident, and two of them are in pretty bad shape and might not make it, but the third one is alert and can probably give us enough information for you to make a composite drawing of the driver of the car that hit them and then took off."

"My head was pounding, and I don't know if it was the bourbon, one of my migraines, or perhaps the bizarre incident of uncontrollably drawing pictures of a woman and a little boy. "At least I won't have to get dressed," I thought. "Okay, Matt, give me about thirty minutes, and I will meet you in the ER. Just be prepared for an unshaven and unkempt sketch artist that probably still reeks of Jack Daniels." Grabbing my sketch pad and the two sketches that had somehow miraculously drawn themselves with the assistance of my right hand, I walked out of my apartment door, took the elevator down to the parking garage, got into my car, and sped away toward Our Lady of Mercy Hospital. "Dear Lady of Mercy, it seems as though you and I have become quite familiar, and I wish this familiarity would stop."

Detective Matt Lelone was at the reception desk in the ER, talking to one of the ER doctors, when he saw me walking through the automatic doors. "My God, will it look like you just came from a homeless shelter. "What happened to you?"

"I'll tell you later, Matt, if I can, as I'm not sure myself."

Matt looked at me with a puzzled expression on his face but then seemed to ignore what I had just said. "We need to go to bed #3 and talk to the only victim who is conscious and able to talk to us. "Get your sketch pad and let's go. Without pausing, he calls out to a nurse, "Nurse, I'm Ashcamp PD's Detective Matt Lelone, and this is our department sketch artist Will Darbin. Please take us to bed three so we can speak to the patient?"

"Follow me, gentlemen," the nurse replies as she leads us past several beds with white sheet-like screens pulled around them, serving both as dividers and for privacy. She pulls one to the side when we reach what would be bed three. "Here you are, detectives. This is Mr. Miller. He has been stabilized and can talk to you." Before she left, she raised the head portion of the bed so he could see us. As she walked by, she gave me a disapproving look.

"She must have noticed the smell of bourbon that is probably still on my person," I thought. Lying on the hospital bed in front of us was a man who was probably 35-40 years old and had both arms heavily bandaged and his left leg in a cast and in traction. "Go ahead, Matt, you can start asking him questions, and when you are finished, I will see if I can get enough information from him so I can draw a composite sketch of the hit-and-run driver.

"What about my wife and son, are they here in the hospital also? Nurse, nurse, can you tell me?"

"The nurse has left Mr. Miller; I'm Detective Lelone from the Ashcamp Police Department, and this is Will Darbin, our department's sketch artist. "We need to ask you some questions. "Do you feel up to answering them?"

Becoming agitated, he replies, "Look at me, how do you think I feel? "Can you tell me anything about my wife and son? Are they alive? Where are they?"

"They are both here in the ER and are being taken care of, Mr. Miller. "I'll check with the nurse and find out their condition and let you know before we leave. In fact, I'll go ask a doctor or nurse now, so why don't you describe whatever you can recall to Mr. Darbin so he can develop a sketch of the hit-and-run driver?" Matt answers.

Taking out my sketch pad, "Please go ahead, Mr. Miller, tell me anything you can remember, no matter how insignificant you think it might be. "We want to catch this person and make sure he or she is held accountable for what they did to your wife, son, and you." For the next several minutes, with the help of Mr. Miller's recall, a definite description of the hit-and-run driver was being formed. Satisfied, I stopped when Matt returned.

"Mr. Miller, I talked to the ER doctors who have been tending to your wife and son, and they told me they were both alive but in very critical condition, and if they survive the next 24 hours, they should make a full recovery. "Would you like me to call anyone for you, your minister, priest, or rabbi, or any of your family members?"

The pain medication the nurse had just injected Mr. Miller with was beginning to take effect, and he was becoming very relaxed and drowsy. Before he drifted off to sleep, he softly mumbled, "Thank you, detective, you don't need to call anyone, as I believe the hospital staff has already made those calls."

As we were walking towards the ER's exit, one of the doctors that Matt had been talking to called out to us, "Detectives, Mrs. Miller has regained consciousness and is calling out for her husband and son. Maybe she will be of some help to you. Don't push her too hard. She's in bed ten, and her little boy is next to her in bed nine.

Walking through the white curtained partition that divided beds nine and ten, I gasped, dropped the sketch pad, strewing its contents onto the floor, and grabbed onto the metal partition, nearly pulling it down.

"Will what's wrong, are you okay?"

Partially regaining my composure, I bent down, picked up the sketch pad and the sketches that had been strewn on the floor, and handed the sketch I had been compelled to draw just last night by someone or something, an unknown enmity that took control of my drawing hand. "I don't know, here you tell me, Matt," as I handed him the sketch of the woman I had uncontrollably drawn.

Now it was Matt's turn to be shocked. "My God, will this is Mrs. Miller, when did you have time to draw this sketch? We've only been here for about an hour."

Ignoring his question, I said, "Matt, we've got to see the Miller's son."

"He's in the partition next to this one Will and if I am right this is what he will look like," I blurted out handing him the picture of the little boy I had been compelled to uncontrollably draw last night as well.

"Will, what's going on you are beginning to frighten the bejesus out of me."

Pulling the curtained partition aside and walking from Mrs. Miller's bed to bed number nine the little boy's face couldn't be seen as he was laying on his side with his back to them. Walking around the bed so they could see what the boy looked like Matt says in an astonished voice, "Holy Mother of God Will it's him, the little boy in your sketch, and he is even missing his two upper front teeth!"

"I don't know what is going on Matt and why now all of a sudden I seem to be possessed by some mystic or psychic ability to see victims before they become one but as in this case the revelation of them were revealed too close to the time they actually did have something happen to them. "I couldn't prevent it as whatever mysterious powers have or are now inside e of me doesn't tell me who, where they are or when something bad will happen to them. "This power isn't a blessing Matt it's a curse!"

"What are you going to do Will, you need some help but I don't have the slightest idea who you can talk to that will even believe you let alone understand why or what is happening to you."

"When we are finished here I am going up to the 6th floor and talk to the counselor I have been seeing during the rehab portion of my treatments since my accident. "I hope they will understand and perhaps come up with some reasonable explanations I don't want them to think I am one of those nut jobs that claims to have been abducted by space aliens and taken aboard their UFO's for experimentation. "Considering what is happening to me right now maybe they aren't nut jobs and something really did happen to them, who am I to judge?"

"Well considering the events of today maybe we both should see your shrink Will. "If you need someone to verify what has just happened here at the ER let me know, I'm here for you buddy. "Here take these sketches of the woman and the little boy you might need them when you talk to your counselor."

"Thanks Matt, I'm sure I will."

Peering at me over the rims of his glasses Dr. Werner Inghart the chief of neurology at of the Our Lady of Mercy Hospital as he put down the note pad he had been writing on during our session he placed the eraser end of the pencil he had using to his lips. His face was expressionless. "Mr. Darbin, while your experiences you have told me about concerning the sketches you say you somehow were made to draw of victims in an auto accident are indeed fascinating as a trained psychologist they don't seem plausible. "Please don't take that statement as one of not believing you I just feel there is some other explanation. "Considering the horrific accident you were in less than a year ago I think the answer or explanation if you will may well be due to some lingering after effects of the life threatening injuries you sustained. "I want to schedule more sessions with you and see if we can get at the root cause of what has suddenly started to happen to you. "In the meantime, I am going to change the dosage in your medication you have been taking; maybe something as simple as this will correct the problem."

"Believe me Doc, it happened and even my detective partner will verify it. "There must be some explanation, and I have to find out and soon. "If I don't I feel that soon our sessions will have to take place in the state mental hospital inside my padded cell."

"Please Mr. Darbin you have my assurances we will find an answer or explanation to the phenomena you have recently started to experience and it won't be from a "padded cell" as you have just described."

Several weeks had gone by, weeks I am relieved to say where no new forced sketch drawing events occurred. "Maybe the change in the dosage of the medication I had been taking did indeed correct the problem. We'll see as this evening I have another session with Dr. Inghart, and he said he wanted me to bring my sketch pad with me; he said something about art therapy," I thought.

Doctor Inghart greeted me as I walked through the door of his office, "Good evening, Mr. Darbin, I see you brought your sketch pad as I requested, good. "Rather than sitting on the sofa, I would like for you to sit at the table in that chair," he pointed.

"Fine, Doc, do you want me to keep my sketch pad with me?"

"Please do, this evening we are going to engage in some art therapy, and you are going to be in a state of hypnosis, and I am going to give you some commands concerning recent crimes and victims, and ask that you sketch a likeness of the perpetrator as well as the victim. "Given the fact that you are aware of these cases, your subconscious mind should take over, allowing you to draw them. "After that exercise, I am going to give you hypnotic suggestions concerning crimes, victims, and criminals that have been committed but you have never heard about them."

"Trick questions or maybe I should say trick suggestions, huh, doc? "Trying to prove or disprove the images I drew of the woman and her little boy, I somehow knew about."

"Well, Mr. Darbin, even though you put it that way, what I really am attempting to accomplish is two-fold, first, as you have just mentioned, if somehow in your subconscious mind you did know about them; and the second, an effort to help you overcome whatever it is you are experiencing."

"Doc, whether you believe me or not, I know they happened and somehow drew both victims in advance and without any prior knowledge of them or the incident. "If I weren't so desperate to find out why this happened to me at this point, I think we are both wasting our time."

"Please, Mr. Darbin, while I understand your frustration and skepticism, trust me at least for this session tonight and then a follow-up visit in a week after I have had a chance to review the results of tonight."

"Okay, fair enough, doc, let's do it," I agreed. Minutes later, I was in the hypnotic trance of Dr. Werner Inghart.

"Mr. Darbin, can you hear me?"

"Yes, doc, I can."

"What is my name and your name?"

"You are Dr. Inghart, and I am Will-- Will Darbin.

"What do you do for a living, Mr. Darbin?"

"I work for the Ashcamp Police Department."

"What is your job at the department?"

"I am employed as one of their sketch artist's."

This kind of dialogue continued for several minutes as Dr. Inghart asked several more questions to build complete confidence between them.

"Okay, Mr. Darbin, I am going to ask you to draw the suspect you drew in the Elka Felice case."

"The image I drew and will now draw is not an accurate portrayal of her attacker, the doctor."

"What do you mean, Mr. Darbin?"

Being in the hypnotic trance I was in, I didn't answer. I drew from memory a replica of the image I had drawn the night I interviewed her at the hospital, the same night of my nearly fatal accident, tore it from my sketch pad, and handed it to him. "Here you are, doc."

Taking the sketch from me, Dr. Inghart thanked me and then asked, "Why isn't the sketch you just gave me an accurate portrayal of her attacker?"

"Doc, that's because she lied to me. "Her true attacker was her spouse or domestic partner, and she didn't want to implicate or identify so he wouldn't be arrested, which was a mistake because not long after that he killed her."

"Putting the sketch that I had just drawn, Dr. Inghart described a suspect who had committed a robbery and assault that happened last week in another city that had gained widespread media attention, but didn't say where it had taken place. The suspect had been apprehended, and his face had been broadcast on several news reports. "Will I want you to draw me a sketch of the suspect that committed this crime. "If you can, in fact, draw the suspect, that will give me a clue that perhaps you somehow did know about the woman and young boy that were involved in the accident. Go ahead, pick up your pencil and begin drawing."

Hearing and understanding his command, I picked up the pencil but could not begin to draw anything, and just sat there for several minutes. "Nothing comes to mind, Dr. Inghart," I tell him.

"Okay, Mr. Darbin, I am going to give you a command, and when I tap my pen on my desk, you will be awakened from your hypnotic trance. I think we're finished for the day."

I awoke when he tapped his pen on his desk. "How did it go, Doc?"

Holding up the sketch I had drawn of the Felice assault suspect, he says, "You had incredibly accurate recall concerning this crime, but when I asked you about a more recent crime that had gained widespread media attention and coverage in Chicago, you couldn't draw that suspect even though he had been shown on television many times."

"Doc, doesn't that prove I am right about what happened to me and the sketches I made of the injured woman and her son? I couldn't have made them up."

"Mr. Darbin, although I still have doubts about that incident, nevertheless, based upon this session tonight, I'm not so sure about them. "I'm going to consult some colleagues of mine who have studied in the field of parapsychology and see if they can explain it. "On your way out, please make an appointment with my assistant for next week, same day and time."

As I got up from the chair at the table where I had been seated I felt the urge and sensation that had come over me the night before Detective Matt Lelone had called me in to meet him at Our Lady of Mercy hospital when I gave him the sketches that I had drawn and that inexplicably turned out to be the images of the injured woman and her little boy. "Doc, I'm getting that strange and unexplainable feeling again," and then sat down and uncontrollably began sketching. After nearly an hour, I handed Dr. Inghart 20 sketches.

"Mr. Darbin, I don't understand or maybe recognize is a better word," as he looked at the twenty pages of sketches I had just handed to him. Thumbing from sketch to sketch, it was easily recognizable that there appeared to be at least 50 or more bodies that had been thrown about on the ground adjacent to what appeared to be a railroad track, a part of which was bent upward away from the ground. Each sketch revealed more bodies burned and twisted passenger cars from a train.

"Doc, I wasn't faking it. Whatever possessed me to start drawing these just took over, and I lost all control of my drawing hand just like I did the night I drew the woman and her little boy. "You just witnessed what happened to me, now you've got to believe me."

"Mr. Darbin, I have to admit something bizarre just occurred, and if I hadn't seen it with my own eyes, I would never have believed you. "I'm going to hold onto these until our next session and we will review them at that time."

A horrible feeling of pending doom or tragedy was starting to take over, "Doc, we have to do something, warn someone but I don't know who."

"You're absolutely correct Mr. Darbin but who do we warn and more importantly what can we do. "If we call the authorities and warn them of an impending catastrophe they will either think we are insane a term I don't like to throw around unless it is absolutely beyond a reasonable doubt, or, he paused, "they will think we are a part of some terrorist plot if in fact such an event does actually happen. "I feel our hands are tied, all we can hope and if you are a religious person is pray the horrific events depicted in your sketches do not come to fruition."

"Now you're beginning to understand how I feel Doc, we're damned if we do and damned if we don't do anything. "I'll see you next week."

A relative uneventful week had taken place and I felt pretty good that there hadn't been any catastrophic train accidents anywhere in the U.S. as I drove to Dr. Inghart's office for our scheduled appointment. The traffic was moderate and I had the news station tuned in on my car radio. "We interrupt the regularly scheduled news hour to bring you a news bulletin. "We take you now to our news bureau in Mumbai, India for a late breaking report of a horrific train wreck that took place a little more than an hour ago in the Indian state of Punjabi. "Early reports say there may be as many as 100 dead and another 400 persons injured as bodies have been scattered on and around the tracks and the train's passenger cars have been twisted like pretzels, many of them burning."

"Oh no, this can't be, when I was obsessed to draw those twenty or so sketches last week neither the Doc or I could recognize the train's passenger cars and this explains why, the accident wasn't going to happen here in America it was going to happen in India." Arriving at Our Lady of Mercy Hospital I hurriedly park my car and then walk rapidly, almost in a jog from my car and through the hospital's main entrance. Impatiently I wait for the elevator to return to the main floor from the fifth. Once the doors of the elevator open I rush inside and forcefully push number 6 which was the floor that Dr. Inghart's office was on. Getting off the elevator I sprint down the hallway to his office and the door is already open. The Doc is with another person, a shrink I speculate. "Doc, I blurt out, "did you hear about the train wreck in India?"

"Yes I did Mr. Darbin, and please calm down. "I have taken the liberty to call a colleague of mine a Dr. Shultzgard, from Munich. "Dr. Shultzgard is an expert in the field of parapsychology, and I feel, based upon what you have experienced with the woman and young boy and what has just happened, I think it has transcended far beyond my training and experiences as a psychologist."

"See, Doc, I told you I wasn't imagining or making up these events."

"I'm sorry, Mr. Darbin, but as I have just said, in all my years and training as a psychologist, I have never experienced such things. "Now, given this latest incident of the horrific train wreck in India, it has been proven beyond any doubt that you are blessed with a psychic ability, something I have never seen or studied before."

"I don't know Doc, it seems more like a curse than a blessing to me."

"Good evening, Herr Darbin. I am Dr. Shultzgard. Please, go over to the sofa and be seated. "I'm most anxious to talk with you. "Dr. Inghart, would you please give me the sketches that Herr Darbin has drawn?"

"Do you want the sketches of the woman and little boy, as well as the train wreck?"

"All of them, please, Dr.Inghart. "Now Herr Darbin, in reviewing your case file, you never had any incidents such as these before-- let's see," he says while thumbing through the papers in his file. "Ah, yes, before the accident that nearly killed you. "It looks like that happened close to a year ago. Is that correct?"

"Yes, Doc, that's right," I replied.

Then, comparing the sketches he had made of the woman and little boy, as well as the ones of the bodies and wreckage of the train, Dr. Shultzgard says as if to himself, "incredible, just incredible, in the forty-plus years I have been studying parapsychology, this is the first time I have ever been able to verify such activity visually."

"Dr. Shultzgard, what do you think, or maybe I should say, do you have any idea what is causing Mr. Darbin to have these experiences?" Dr. Inghart asks.

"I don't know if we will ever know the exact reasons, Werner, but it is my educated guess that the head trauma Mr. Darbin suffered in his accident has something to do with them. "In parapsychological study, it is the study of phenomena suggesting that the assumption of a strict separation between subjective and objective may be wrong. Human experiences, such as Herr Darbin's, suggest that some phenomena occasionally fall between the cracks and are neither purely subjective nor purely objective. " From a scientific perspective, such phenomena are called anomalous because they are difficult to explain within current scientific models. "Now, as a result of Herr Darbin's experiences I believe we at last have such a scientific model. "The anomalies I am talking about fall into three general categories, ESP, PK or Psychokinesis or direct mental interactions with physical objects animate or inanimate as well as Bio-PK which is direct mental interactions with living systems. "Also, from his experiences, he seems to possibly fall within the category or ability of having automatic writing or an entity that guides the hand of a living person. "Herr Darbin, how would you like to go back to Munich with me so we can study your incredible talent under a controlled laboratory environment?"

"I don't know, Doc. I've already missed a lot of work because of my accident, and I'm not sure the Ashcamp PD will give me more time off, but I would like to learn at least how to control what happens to me. "I'm not sure how long I can keep my sanity if I don't do something."

"Excellent, Herr Darbin, I will speak to your employer and make all the arrangements, and don't worry about money, as my university and the German Government will highly compensate you. "Meet me here at Dr. Inghart's office three days from now, and we will make final preparations for our trip to Germany."

The events of the day had been exhausting, and on my way home, I stopped at the Ashcamp Police Department and told my good friend, Detective Matt Lelone, about my latest sketching incidents and that I was going to take a leave of absence and go to Germany with Dr. Shultzgard. "Can you believe it, Matt? I'm going to Germany and become an experimental lab rat?"

It's about time Willy the Sketch Artist did something valuable and noticeable around here, and when you become famous and in all the medical journals, I hope you will still have time for lowly buddies here at Ashcamp PD."

I was exhausted when I finally walked through the door of my apartment, but I was too wound up to go directly to bed, so I poured myself a very tall glass of Jack Daniels, good old Tennessee Sour Mash, sat down, and took a long and healthy drink. As I began to relax and started to get drowsy, the entity, I think, that Dr. Shultzgard had called it once more, took control of my right drawing hand, and I started sketching.

Only this time, I did it more feverishly than I had ever done since I started experiencing the blessing as the good doctor had called it. "Sorry, Doc, it's more like a curse, not a blessing." By now, the whisky and the pain medication were really kicking in, and I lay my head down on the desk and passed out, not bothering to look at the sketch I had just drawn.

Friday had arrived, and I was set to meet Dr. Shultzgard, but the meeting never took place.

"Will, you here?" Detective Matt Lelone called out after he had been let into my apartment by the apartment manager. Walking over to my desk, where I was still seated, slumped over, he could see the empty whisky glass, the spilled bottle of Jack Daniels, the bottle of pain medication, and a sketch that was directly under my face he could see that the augur sketch artist had drawn his last sketch a picture perfect likeness of himself his latest but last victim.

The End

Impossible, it Couldn't Happen— —Not in My Neighborhood

A Short Story
by Ralph L Myers

Impossible, It Couldn't Happen, Not in Our Neighborhood

Taking a sip from his morning coffee and then putting it down on the kitchen table, Jeff Arthur picks up the newspaper and hands it to his wife. "Honey, just look at this. There were over five murders that took place right here in Jasonville. "According to the story, there have been more than 25 homicides committed so far this year.

Taking the newspaper being handed to her and glancing at the story Jeff was referring to, then putting it down on the table, she questions almost in an apathetic voice, "So, what else is new? It's tragic, but it doesn't affect us. "Besides, look at where those murders took place over in the bad part of town. "Murders and violent crimes are or seem to be a way of life there. "I sure am glad we live in this safe, low-crime neighborhood. Stuff like that never happens here."

Getting up from his chair, walking over to the coffee pot, and pouring another cup of coffee, taking in what his wife had just said, "You're probably right, Justine, nothing like that has ever happened here in this neighborhood. I just guess I am overreacting, worrying about something I need not be overly concerned about. "I do feel bad for those poor people who were murdered and their families. I just can't imagine what they are going through right now."

Again, in a dismissive sounding voice, "It is sad, Jeff, but who knows what kind of people they were, gang members, broken homes, criminals? "I doubt if they were upstanding citizens."

"My god Honey, you make it sound as though they deserved to have happened to them what did happen, how can you say such a thing? "We don't know anything about the victims or what kind of people they were."

"For god's sake, Jeff, give me a break! "Maybe you're right, but what happened has happened, and there isn't anything either of us can do about it. "Let's just drop the subject. I don't want to talk about it anymore; it's depressing. "Besides, the police will catch those who are responsible, and the criminal justice system will put them away. "I've got a full day ahead of me and have more important things to take care of or think about, which reminds me I have to be at Cynthia's house on Flora Drive, a few blocks away, to help plan her daughter Melanie's 18th birthday party in less than an hour from now. "Please go and wake up John, he is supposed to help us."

A sleepy-eyed young man slowly walks into the kitchen. "Mom, Dad, what's for breakfast?"

"From here, nothing, hurry up and get dressed, you're going over to Melanie's house with me and help decorate for her birthday party, we can stop along the way and grab something to eat."

"Mom, that party isn't until tomorrow night. What's the rush? "Besides, I was planning on going surfing with the guys this morning."

"Jeff, help me here."

"Forget about it, John, just call your buddies and tell them you'll have to make it later after you help your Mom."

Taking half a gallon of milk from the refrigerator and gulping most of it down directly from the carton and then putting it back on the shelf, "Mom, can't you do this without me? There are some awesome waves at Zuma right now."

"No, just hurry up, John, get dressed, I'm sure there will be more awesome waves tomorrow. "We need to leave in ten minutes."

The black BMW 325i with darkly tinted windows stops, blocking the bicycle lane. One of the occupants, a young man, probably 16 or 17, gets out of the car and walks over to the telephone pole next to it, smiling. He pulls a flyer that was thumb-tacked to it. Getting back into the car and turning to the other three guys inside, "Looks like we're going to a party tomorrow night, dudes."

"Radical dude, I know that chick Melanie, she goes to my school. "I would like to go out with her sometime."

"There's no way, man, that chick wouldn't have anything to do with you," another occupant of the BMW chimes in.

"We'll see, dude, we'll see. "I am surprised there's a flyer advertising this party I wouldn't have thought they were inviting everyone."

"They didn't man, some other chick at school heard about it, and I guess she wasn't invited, so in spite of it she made a bunch of flyers at Kinko's and posted them all around the hood."

"Wow, she sure Gomer Pyle'd them."

"Gomer Pyle'd them, what is that supposed to mean?"

"Where have you been, man, you know, Gomer Pyle the dumb Marine on that old T.V. show that was always doing stupid stuff and then saying Surprise, Surprise!"

"I never saw the show, so I still don't know what you're talking about, dude."

"Well, Melanie, as Gomer Pyle always said, surprise, surprise, we'll see you tomorrow night. "I hope you have plenty of beer as twenty more of us will be showing up."

It was a warm but not too hot summer night. A cool ocean breeze was blowing onshore as John parked his four-wheeled vehicle on the street just outside the gated entrance to Melanie's house. With two six packs in hand, he was greeted by several of his surfing buddies who had already arrived. Hey John, toss me one of your brews, two of his buddies called out. John and most of his buddies were over 21, and he was careful that none of the younger partygoers got hold of any of the beers he was happy to share. Walking inside the house, he greeted and hugged the birthday girl, who was seated by four of his best friends, Adam, Jonas, Brett, and Richard. He gave each of them one of his beers.

"Hey John, let's get out of here. There are too many teenagers at this party. "Let's go shoot some pool at Billiard Heaven," one of them yelled so he could be heard in the noisy room.

"Nope, you guys go ahead, I want to stick around for a while, maybe I'll join you later."

"I'm going to stay here too," Jonas says.

At that very moment, the black BMW 325i and three other cars pulled up and parked on the street outside of the yard where the birthday party was taking place. There were about 15 to 20 of them, girls and guys who got out of the cars. Walking together as a group, they approached the gate that led to the backyard entrance, where the party was taking place. They were greeted at the gate by two partygoers. "I'm sorry, this is a private party, and you can't come in unless you have an invitation."

One of the guys who appeared to be a leader pulls out the flyer that they had taken down from a telephone pole the day before and hands it to the partygoers who had stopped them. "Here's our invitation," he said as he handed them the flyer.

"This can't be right. Melanie didn't have any flyers printed for her party."

"Well, we have one, so somebody must have put this on the telephone pole I took it down from. "So, get out of our way, we're going in."

Two or three other partygoers inside the yard noticed what was going on and walked over to help the other two, who were trying to turn the party crashers away. "C'mon, guys, you can't just walk in here and ruin this party, it's private, so please just leave."

Two of the male party crashers walk away and go back to the BMW and get inside. Moments later, they reappear, one with a handgun, the other with a metal club. The one with the gun fires it up in the air two or three times while the other proceeds to hit one of the young men who had confronted them and were trying to prevent them from entering the party.

Hearing the gunfire and the commotion taking place, John walks out of the house into the backyard towards the gate to make sure no one was bothering his four-wheel vehicle. By this time, the person with the gun leveled it and fired indiscriminately into the crowd of partygoers. Two of the bullets hit John in the upper torso and chest. Bleeding profusely, John manages to stagger back into the house and collapses on the floor. His good friends Jonas and Richard try to comfort him as he begs for help. They stand by helplessly as their good friend John dies in their arms.

It had been a long day, and John's mother and father decided to retire for the night. Both quickly fell asleep and slept soundly through the night. It was around 7:30 am the next morning when John's father, Jeff, was awakened by the ringing of the front doorbell. Walking by John's room but not looking in, he thought, "*must be some of his buddies coming to pick him up to go surfing.*" Glancing through the peephole, he saw two men dressed in suits that he did not recognize. Opening the door, he asked, "Can I help you?"

They identified themselves as detectives from the Jasonville Police Department. "Do you know John Arthur?"

"Yes, he is my son. Is he in some kind of trouble?"

"We are sorry to inform you, Mr. Arthur, but your son John was shot and killed at a party near here last night."

Grasping hold of the open door to steady himself, his mind raced back to yesterday morning as he and his wife were casually talking about the murders and violence that had been taking place in Jasonville, *"but in bad areas of the city, not in their neighborhood.* "Please come in, officers. I have to get John's mother and tell her what has happened; she is still asleep."

"We'll wait here," the detectives tell him.

Going to their bedroom, Jeff now had the duty of waking his wife and telling her what had happened; their only son had been murdered. "Honey, wake up, something horrible has happened."

Sitting up and rubbing the sleep from her eyes, "What, Jeff? What has happened?"

Tears welling in his eyes, "It's John, as he struggled to get the words out of his mouth.

"John, what about John!"

"Honey, he's dead, he's been murdered!"

Instantly, screams of anguish and pain came from John's mother, the woman who had conceived him, carried and nurtured him in her body for nine months, and then lovingly held him in her arms just after the moment of birth. "No, she screamed, it can't be, something like this doesn't happen to a family like us or in a neighborhood like the one we live in. You and the detectives must be wrong, it couldn't have happened, it's impossible!"

Now the Arthur family had suffered the worst kind of loss imaginable, the murder of their child, and were cast into the American criminal justice system as victims, and as Jeff had wondered just yesterday morning, he now not only would imagine what the other families are going through, he would now be forced to find out for himself.

After more than four months, an arrest was made of the person who had murdered their son. During that time, they had been in constant contact with the police department and detectives assigned to their son's murder case, as well as the deputy district attorney who was assigned to prosecute the murderer.

John's mother thought back to the morning before he had been murdered, when she told her husband the criminal justice system would put murderers away. Now they were meeting with the prosecutor and were in for a rude awakening and education about the criminal justice system as it pertained to the victims and the judicial process.

"Mr. & Mrs. Arthur, the case against your son's murderer is not as strong as it should be, and we are considering offering a plea agreement to the defendant."

"Wait a minute, are you talking about a plea bargain? "My wife and I don't want this murderer to be offered any such bargain, agreement, or whatever it is called."

"Mr. Arthur, while I understand your concern, you nevertheless don't have the right to prevent us from doing so, as now the people of the state of California are the plaintiffs in the case. "I only feel I, as the prosecutor, have the moral duty to advise you of the possibility we will offer a plea agreement, not a legal obligation."

"This guy murdered our son, and I want him held accountable and no slack given him."

"Again, Mr. Arthur, I understand and we want to convict him as much as you want him to be but I think you and his mother will agree if by offering him a plea agreement of reducing the charges from first to second degree murder and he will admit his guilt and accept the agreement you, we will be assured of a conviction and he will indeed be held accountable. "If we don't or he won't accept the offer, and this case goes to trial, there is always the chance he will be found guilty of even a lesser charge than second-degree murder, or possibly acquitted of the charges and walk away a free man."

My wife and I had just received our first lesson about the American Criminal Justice system, and so far, it was a rude awakening. We were our murdered son's advocates; we were his voice since he had been silenced forever.

No longer did we watch the news or read the newspaper's accounting of yet another murder, another family being victimized in an apathetic, nearly unsympathetic manner; we had to be active in the process. The only way we could get through, survive what lay ahead for us, was to be involved every step of the way.

Fast forward, eighteen months since the murder of our son. As I have written, my wife and I felt it absolutely critical that we be involved in every hearing, court appearance, and meeting with the district attorney's and homicide detectives' offices. We had attended and made our presence known in the juvenile court system as the murderer was under the age of 18 when he killed our son, then on to the municipal and superior courts for hearings, as he was found unfit to stand trial as a juvenile. Then, finally, the preliminary hearing was held, where it was determined there was enough evidence to show he should be tried (as an adult) for John's murder. During this time, we were repeatedly made painfully aware that victims in the American Criminal Justice System don't have any constitutional rights in the federal constitution, whereas the criminal has the following.

Right to Counsel (at no cost if they can't afford it)
Right to Due Process
Right to a Speedy Trial
Prohibition against Double Jeopardy
Prohibition against Self-Incrimination
Protection against Unreasonable Searches and Seizures
Rights to have warrants issued only upon probable cause
Right to a Jury of Peers
Right to be Informed of Accusations
Right to Confront Witnesses
Prohibition against Excessive Bail
Right to a Grand Jury Indictment
Prohibition against Excessive Fines
Prohibition against Cruel and Unusual Punishments

Now, at long last, the murderer would stand trial, and we prayed that he would be found guilty and held accountable for murdering our son. Once more, we met with the deputy district attorney about it.

As we waited for the deputy district attorney in the reception area outside of his office, we glanced at the walls that were adorned with replicas of the Bill of Rights, the U.S. Constitution as well as pictures of the Governor, President of the United States, the U.S. Attorney General, and the District Attorney of the county where the trial was being held. All these artifacts were strategically placed on the wall surrounding a floor-to-ceiling painting of the blindfolded Lady Justice carrying a sword in one hand and the scales of justice in the other. Finally, we were summoned to enter the deputy district attorney's office.

By now, we were on a first-name basis with the prosecutor who would prosecute this case for the people of the State of California. "Good morning, Jeff and Justine, well, it looks like the day we have all been working for and anticipating is about to begin. "I just want to go over a few ground rules with you when the trial begins. "First, no matter what is being said by the judge, defense attorney, me, or the witnesses, you must not speak out or show any emotions. "If you do, the judge may have you removed from the court and barred from returning any more throughout the duration of the trial. "It could be considered prejudicial to the defendant receiving a fair and impartial trial by perhaps influencing members of the jury."

"We understand, we don't like it, but we understand."

Continuing, the deputy district attorney says, "During testimony, the coroner will be called to the stand to explain the injuries sustained to your son's body and the cause of death, and there will also be crime scene photos, both graphic and gruesome."

"Honey, I don't think I can stand to see those pictures, as I want to remember an image of John as being the fun-loving, youthful, and good-looking son that he was. "If I see the crime scene photos, I feel those images will be branded into my memory for the rest of my life, and I will always remember him like that instead."

"That's perfectly understandable, Jeff, and I will warn you in advance, and you can excuse yourself and wait outside the courtroom until the coroner's testimony is finished."

"Guess we don't want to spoil that alleged murderer's chance at receiving a fair trial, do we?" Jeff says sarcastically.

"Certainly not, Jeff, and we most assuredly don't want the judge to declare a mistrial because he feels the jury has been improperly influenced by an emotional outbreak of you or your wife or anyone else in the courtroom."

On the drive home, Jeff and his wife, Justine, talked about the meeting that had just taken place with the deputy district attorney. "This brings home once more that the victims have no rights or standing in the criminal justice system. "This needs to be changed, and it looks like the only way it can be is through an amendment to the U.S. Constitution giving victims a standing in it."

Two decades have gone by since that horrible summer night when our son was murdered and countless others have been murdered in America and forced to travel the road of the criminal justice system as we were, and yet there are no rights afforded to the victims in the U.S. Constitution, even though there have been attempts to right this injustice. For us, John's parents, we are still dedicated to fighting for victims' rights, but as we get older, we're not sure if that will ever happen in our lifetime. But we won't give up. We consider our fight to be perhaps our final act of parenting for our John, as I'm sure those of you out there who are dedicated and caring parents, you never stop being Dad or Mom, and you are only doing what you feel is right for your child, even though they are no longer with you.

In the years that have gone by and I trying to educate others about the plight of crime victims, I have arrived, perhaps biased, at the conclusion that society in general, particularly those who have never suffered the loss of a loved one to murder, don't feel the need for victims to have rights in the criminal justice system. I only hope and pray they will never be cast into the position of being a crime victim or a survivor of a murdered loved one. Also, many naively still think that bad things only happen in the bad parts of town and could never happen in their safe, secure neighborhood. Sadly, it does, it can, and will happen. Now, as I sit at the breakfast table in the morning, reading the newspaper, I think back to the conversation my wife and I had the morning before John was murdered. One morning, I wrote the following.

"Impossible--not in my Neighborhood"

With a cup of coffee and the daily newspaper in hand, another night of senseless violence and wanton criminal deeds is revealed.

My sense of outrage, sadness, and empathy for the victims is lessened, nearly deadened by the distance between where I live and where horrific acts were committed.

Thank God, if those things must happen, they occur elsewhere and surely could not take place in my safe neighborhood.

The reward of my hard work has manifested, allowing me to live in a community where crime is nearly nonexistent. I feel secure and safe in the knowledge that my children and their friends are decent and good.

Days, weeks and months swiftly become years. Time passes by as the enigma of violent crime continues to escalate.

Our neighborhood is still perceived to be safe it seems to be an island surrounded by gangs and crime.

Urgent action must be taken while there is still time.

New victims are added to the statistics as those figures grow ever larger.

Yet another mother weeps when her son or daughter is added to the toll.

Once more, as countless times before, we bid farewell to our loving, treasured child.

Cautioning him or her to be careful and good, avoiding dangerous places and peers, both bad and wild.

As they abide by our wishes and follow our advice, they innocently attend a party, a movie, or a function close to home.

Our whole world is shattered, its course forever altered by the knock of a stranger at our door, or an unfamiliar voice on the phone.

Mr. and Mrs. Safe, secure citizen, the impossible has happened. I am sad to advise.

Your son or daughter died this night.

A victim of violence, of a senseless murder, without reason or logic, I am sorry to say, and I can't begin to explain why.

While reality begins to set in, as family, friends, relatives, and neighbors stricken by disbelief, overwhelmed by sorrow and grief, ponder what has happened, struggling to grasp why it did or how it could.

One familiar thought is painfully real. What has happened to my child could not be. Or so I thought and truly believed. IT IS IMPOSSIBLE, IT DOESN" T HAPPEN IN MY NEIGHBORHOOD.

THE END

Detective Atessa And The Case of The Hijacked Holidays

A Short Story

by Ralph L Myers, AKA
Grandpa Ralph

Detective Atessa and the Case of the Hijacked Holidays

By Grandpa Ralph

It was early Christmas morning as Santa Claus guided his team of eight reindeers led by Rudolph gliding into their North Pole landing site inside the secret sleigh hangar that was camouflaged by the purest and brilliantly white snow. If someone was near the hangar that didn't belong there, it was so expertly concealed that they would probably go right by it and think it was yet another large snow drift, as the entry to the site was impossible to detect, even from the air. However, just to be on the safe side, Santa had huge, ferocious polar bears at his command, stationed all around the hangar and guarding it closely in case, by some miraculous accident, an intruder discovered it. If somebody accidentally found the hangar, they wouldn't stick around long when they were chased away by the polar bears. Thousands of Christmas's had come and gone and yet the reindeer sleigh hangar had never even been seen by anyone other than Santa Claus and his reindeers, not even Mrs. Claus or any of Santa's elves knew of its existence, they only were aware that he parked his sleigh somewhere and on Christmas Eve they would see it parked outside his toy shop as his elf toymakers would load it with toys for all the children in the world.

After parking the sleigh in the hangar, Santa unhitched the reindeer and led them outside. Until next Christmas Eve, the reindeer were free to roam around Santa land and were allowed to explore the frozen forest and ice skate on the magical pond that stayed frozen year-round, even when it was summer and the sun would shine past midnight. Ice skating, using only their reindeer hooves, which served as their skates and were better than the normal skates humans and Santa's elves used. In fact, skating on the year-round frozen pond often helped keep the reindeer's hooves strong, and the constant skating helped keep them physically fit, so they would be ready for the daunting task of pulling Santa's sleigh around the world next Christmas Eve.

Leaving the hangar, Santa greeted the Polar Bears that gathered around to welcome him and the reindeer back to the North Pole. Playfully, they frolicked in the frozen white snow as they welcomed Santa back. It was hard to imagine that the gleeful Polar Bears were ferocious and would ward off any intruder, even if it meant they could lose their lives. But they knew who this jolly old man with the flowing white beard and hair, wearing a bright red suit that really was a stark and brilliant-colored contrast to the pure white snow, was. Also, the Polar Bears knew all of Santa's elves and Mrs. Claus and would not try to chase them away from the reindeer hangar they so carefully guarded. This was particularly good now as Santa was just returning from his annual flight around the world and soon the head elf Myron arrived and looked puzzled as to why Santa returned here to this giant snow drift each year and would summon him to bring his snowmobile here and give him a ride back to his home which was several miles away, and now in the cold dark winter morning they would be guided back to Santa's home in Santa Village by the multi-colored northern lights that appeared in the sky above this time of year.

Santa wondered as he sat on the back of the speeding snowmobile being driven by his chief elf Myron if people around the world ever thought or even realized that due to different time zones and the international dateline Santa's annual Christmas Eve journey was in reality two days, for when it was December 23 in one part of the world it was already December 24 in the other part of the globe. So, in order to deliver toys to all the good little girls and boys on December 24 Santa would start his journey on the day before Christmas Eve and fly to the part of the world where it was in fact already Christmas Eve and by the time he finished his deliveries to where it was already December 24 he would head back to the other part of the hemisphere where it would soon be December 24 as well. The thought of it left Santa exhausted, since he had made deliveries two days in a row, even though December 24 is only one day.

Soon, Santa's thoughts and tiredness disappeared as the snowmobile sped out of the frozen forest and started the downward direction into the valley below. He always loved this part of his return home as the bright lights of Santa's Village illuminated the frigid, dark night like a sparkling diamond from afar. "In just a few minutes, I will be home and be greeted by Mrs. Claus and all my hard-working, dedicated, and loyal elves. *"Just like in the story of The Night before Christmas, I will be ready for a hard-earned and much-deserved 'long winter's nap,'* he thought.

Finally, on arriving at his home in Santa's Village, a strange feeling came over Santa as Mrs. Claus and his faithful and loyal elves didn't come running outside to welcome him back home. *"That's odd, where is everyone, especially Mrs. Claus? "Something must be dreadfully wrong,"* he thought as he opened the front door to his home upon return from his annual Christmas run, walking inside. Once he entered his North Pole home, he saw Mrs. Claus and the elves he left in charge were intently examining the calendar for the coming year that would begin less than a week away.

"Oh dear," Mrs. Claus exclaimed, "there is something drastically wrong with the calendar for next year; none of the holidays are shown. "How could this be possible?"

As Santa entered the room, Mrs. Claus looked up from the calendar she was carefully examining with the elves and saw Santa Claus. "Oh, Santa, I am so happy to see you. Then, frowning, she again looked down at the calendar for next year. "The elves and I have a mystery on our hands which I am sure will puzzle you as well, Santa."

"What is it, my dear Mrs. Claus? Surely there must be a simple explanation as to why the new calendar for the coming year doesn't show the holidays. "The printer must have made an error somehow when they printed the calendars and did not list the dates or days of the holidays."

"Santa, I would like to think that is the reason why there aren't any holidays listed on the new calendar, but I have checked with other calendar makers, and they have told me that the holiday dates and days have disappeared from their calendars as well," one of the elves lamented.

Myron, Santa's chief elf, looking across the room where Mrs. Claus and the other elves were examining the new calendar that didn't have any of the holidays listed for the coming new year, noticed the bright red light was flashing on Santa's personal and secure telephone. Walking over to the phone and picking it up, he quickly recognized messages had been left by several of Santa's friends and holiday colleagues. "Santa, you had better come over here and review the messages on your voice mail."

Walking over to the phone and pressing the "play message" button, his voice mail inbox was full. When he pressed the button, a Christmas tinkling bell could be heard, and then the sound of the person who had left the message could be heard. "Santa, this is Old Father Time, and since the New Year is just a week away, baby New Year hasn't arrived as he usually does on Christmas day, and unless he gets here soon, I won't be able to go over his duties of presiding over a new year. Then Old Father Time's tired voice said, "Santa, this past year has been exhausting, and I hate to admit it, but I am getting too old for this kind of pressure. "Please give me a call so we can discuss this."

As soon as Old Father Time hung up, a new message came on. "Santa, this is Cupid calling, and usually on Christmas day, I receive my love arrows from Saint Valentine, and they haven't arrived yet. "Without those arrows, I will be unable to cause love to be spread around the world on St. Valentine's Day, and without love, the whole world will be cast into chaos."

When Cupid hung up, the voice of a leprechaun with a very heavy Irish accent could be heard on the voicemail. "Santa, me lad, this is Clancy, St. Patty's leprechaun, and he wanted me to call you to let you know we haven't received our annual shipment of four-leaf clovers and shillelaghs. "Without them, Santa, the entire world will be deprived of the 'Luck of the Irish' and will be overrun by evil leprechauns, and the snakes he lured away will return to Ireland if there isn't a St. Patrick's Day next year."

Santa then played message after message from the spirits of other holidays, all bearing similar warnings that their particular holiday would be left off next year's calendar.

"Oh my," Santa said as he placed the phone receiver back onto the telephone. "We have a problem of catastrophic consequences so serious that the world will be forever changed for the worse if even only one year goes by without any of those holidays. This calls for drastic action. "Myron, summon a meeting of the WCOH (World Council of Holidays) to meet here at the North Pole, not later than two days from now, as time is getting very short before next year begins, and something must be done immediately."

"Right away, Santa, I will send Rudolph into orbit around the globe as a satellite, calling upon all member countries of the WCOH to send their representatives to this holiday-saving meeting," Myron said.

Even though the world was experiencing a pending no-holiday crisis, Santa laughed and suggested to Myron, "Rudolph is probably as exhausted as the rest of the reindeer, so why don't you Tweet, email, or send out Facebook messages instead of putting him into orbit?"

"You're right, Santa, that's what I will do," Myron replied.

NORAD Headquarters, (North American Aerospace Defense Command, Cheyenne Mountain, Colorado)

Seated at his command post deep inside NORAD headquarters in Cheyenne Mountain, General Monty Blanc is handed a report marked ***Top Secret*** by an intelligence analyst. "General, our radar sites are monitoring hundreds of what are being reported as unidentified aircraft travelling in the direction of the North Pole.

After the General takes the Top-Secret report from the analyst and quickly reviews it, the General tells the security analyst, "If the entire world knew this, they would have definitive proof that Santa Claus does indeed exist." My staff has been briefed by the President and the CIA that a high-level emergency meeting of the World Council of Holidays has been called by old St. Nick himself. "But because of the highly classified nature of this report, no one will ever know about it, nor will it ever be acknowledged by our government, as well as those other international governments. "Sort of like UFOs and the existence of Area 51," the general says.

Meanwhile, in Santa's Village at the North Pole, which is 3171.7 miles away from Cheyenne Mountain in Colorado, Santa is calling to order a meeting of the World Council of Holidays to discuss and come up with a solution to the world's problems, as there are no holidays in the soon-to-be new year.

Rapping a candy cane-shaped gavel at his place on the podium, where he (Santa) presides as the Chairman of the World Council of Holidays, the meeting is called to order. "Ladies and gentlemen, this meeting will now come to order, and as always, the official international Holiday Spirit Language will be spoken at this meeting, so no interpreters will be required."

"Here, Santa," his chief elf, Myron, says. "This is the official agenda of this meeting."

Taking the agenda from Myron and looking it over, and then looking out at the hundreds of people attending this meeting, Santa says, "As each of you will note, there is only one item of discussion on the agenda each of you has been given a copy of. "That topic being the pending world crisis of no holidays being observed or celebrated in the coming new year, which is now less than a week away, and what we can do about it."

Although everyone in attendance had been briefed on why this emergency meeting of the WCOH had been called, there were nevertheless murmurs of concern and fear. "Santa, what is the world going to do if anything can be done in such a very short time before the new year begins?" someone calls out from the floor of the meeting room.

"Yes, it appears like we need some kind of a miracle," another person calls out.

Trying to calm those in attendance and ease their fears, Santa tells them, "I have a proposal and more than likely a solution to the impending holiday crisis here with me today. Then, turning and motioning to a beautiful young girl seated on the podium with Mrs. Claus, Myron, the chief elf, and other members of the WCOH's executive committee, Santa continues. "Ladies and gentlemen, I want to introduce to you a young lady whom I have the utmost confidence in and feel without hesitation will be able to solve the new holiday crisis, a detective who specializes in holiday mysteries, Atessa. "Although she is very young, I can assure all of you here today that she is wise beyond her years and has a lot of experience in solving what always appear or seem to be unsolvable problems. "Atessa, would you please address the esteemed members of the WCOH?"

Walking to the podium, Atessa smiles and hugs Santa Claus. "Thank you, Santa, what an honor it is for me to have been invited to these very important and crucial meetings of the World Council of Holidays. "Even though I am only nine years old I take this responsibility and opportunity being cast upon me to use my detective agency AK Detective Agency Ltd. or Atessa Keyvan Detective Agency very seriously, and I promise, ladies and gentlemen, I along with my able and expert assistant Keyvan will solve this no holidays on the calendar catastrophe. My detective agency was made aware of this mystery on Christmas Eve when an article appeared in the Journal of Bah Humbug magazine which stated that a nefarious and evil disciple of the ban all holiday's movement who goes by the name of Xmas Repeal was quoted when he said "due to the season of joy and cheer which almost everyone worldwide acknowledges the Christmas holiday personifies it would be impossible to cancel it. "Mr. Repeal then mentions to the reporter covering the story that he has developed a plan that will not only end Christmas but also all the known holidays celebrated all over the world by different cultures and customs. When the reporter pressed him on what the plan was, his only response was "the whole world will find out in just a few days, let's just say the new year that will soon be upon us will usher in what I like to think of as a 'Holidoomsday."

"Oh, how horrible," a person in the audience cried out. "Detective Atessa, do you and your assistant Keyvan have any idea or clues as to what the evil Xmas Repeal meant?"

"At the time he made that claim I didn't have the slightest clue or idea what he meant but given the information passed on to me from Mrs. Claus about the calendar dilemma I think the world now knows, based upon Xmas Repeal's statement to the reporter what his evil plan is and my assistant Keyvan and I will be working day and night on our own plan to thwart Xmas Repeal's evil plan. "Now, Santa if there aren't any more questions, I would like to be excused from this meeting so Keyvan and I can save not only Christmas but also all the other holidays that are celebrated every year which I can truthfully say would also be every person on the earth's birthday since those too are considered a holiday and milestone in their lives."

Moans could be heard from everyone attending the meeting when the subject of no more birthdays was brought up by Detective Atessa. "No more birthdays, what will happen to the bakeries in the world if they don't have a reason to bake birthday cakes?"

"Yeah, and what about the companies that make wrapping paper so that birthday presents can be wrapped?"

"And don't forget about the Christmas presents," another person shouted. "Please, Detective Atessa, leave right away and do what you can to put a stop to the evil Xmas Repeal's diabolical plans."

"Thank you, everyone, for your support and confidence in Keyvan and me, and I promise I will save Christmas, all the other holidays, and birthday celebrations worldwide," Atessa says as she hurriedly leaves the speaker's podium.

"Thank you, Atessa and Keyvan, too, from the bottom of my heart, as well as those here at this WCOH meeting and billions of unsuspecting people worldwide," Santa says as Atessa walks out the door. "What you are about to do will rank among the highest, most noble achievements in the history of the world, and like Santa Claus and Christmas, your names will be remembered for centuries to come as the detectives that saved all holidays and birthdays. It was apparent that Santa was very worried, as he wished Atessa well and did not send her off with his traditional, jovial, jolly "ho, ho, ho" because it usually made his belly shake like a bowl full of jelly, and Santa thought this might make him ill.

Soon, Atessa was on board a private North Pole Airlines Charter flight heading back to the Province of British Columbia, Canada, where her detective agency's office was located. When the private jet landed at the Vancouver, BC International airport, Atessa was greeted by her assistant Keyvan and a contingent of Royal Canadian Mounted Police who escorted her directly to her detective office, AK Detective Agency LTD, with emergency lights flashing and sirens blaring. While en route to her detective agency, Atessa asked her assistant, Keyvan, if there had been any new developments regarding the ongoing crisis of no holidays appearing on the New Year calendars.

"The only odd thing that has recently happened that was not made public by Xmas Repeal is he anonymously met with the Chairman of The International Society of Calendar Publishers, Mr. Monthday, at a secret location in the Yukon," Keyvan tells Atessa.

"How did you find that out, Keyvan, and is that meeting verifiable?"

"While you were at the North Pole, I was with Mama and Grandma Francine at a Hallmark Store and the manager of that store was very angry because the supply of calendars for next year had arrived and upon inspection by her she thought the calendar's printing company they had ordered them from had made a huge mistake by leaving all the holidays off every calendar. "Also, she questioned why all the calendars she had just received had been sent from Yukon, Alaska, instead of their corporate office in Kansas City, Missouri?

"That's very strange, Keyvan; did the Hallmark store manager say anything else that seemed unusual or strange?"

"Only that the cartons containing the new but incorrect calendars she had just received had a signed letter from the chairman of the International Society of calendar publishers, Mr. Monthday and Mr. X Repeal that from this year forward all calendars would be shipped by them as had been agreed upon with him and Mr. Repeal at their just concluded meeting in the Yukon."

"So, Keyvan, that letter would indeed verify the secret meeting between Mr. Monthday and Xmas Repeal, and the two of them are in collusion in creating what Mr. Repeal has referred to as a Holidoomsday, and I bet the rest of the board of directors at the society of calendar publishers were totally unaware of this devious plan. "As a result of their plotting to eliminate all the holidays, they have given me a path to subvert their holidays, and I will commence operation Holiday Salvation immediately, and as my operation Holiday Salvation progresses, Santa will be advised and also be involved in many of the plans of attack," Atessa replies.

What are we going to do?" Keyvan asks.

"There are several courses of action we will need to take, Keyvan. "First I will ask Santa Claus to personally send an encrypted message signed by him to all the members of the board of directors of The Society of Calendar Publishers advising them of the plot that their director Mr. Monthday and Mr. Xmas Repeal have secretly put into action asking that they convene an emergency conference call or a GO TO computer meeting to fire the Chairman Mr. Monthday. "At the same time, during the meeting, Santa will ask them to have a motion passed reversing and voiding any agreements they have, whether they are secret or known, with the Society of Calendar Publishers. "Once this is done, the directors will pass a motion to notify all publishers of calendars not to sell or distribute any of the calendars that don't show the holidays on them and recall them from any business enterprise that sells or offers free calendars on a promotional basis. "This first step in my plan to save Christmas and all the holidays that are observed each year, particularly with the help and clout of Santa Claus, will stop this devious plan from happening."

"Okay, Atessa, once this first step has been taken, what about all the calendars that have been distributed that don't show the holidays on them?" Keyvan asks.

"Although it will be a daunting task, Keyvan, it is not impossible to get the message out worldwide that next year and all the years after that, people everywhere can still celebrate the holidays they choose to and their birthdays as well. "Again, using Santa Claus and the power he has he will ask that all governments of the world let the citizens of their countries know that none of the holidays have been cancelled through a meeting of the Security Council at the United Nations and that they should destroy any calendars that don't have the holidays shown on them and new calendars containing the holidays will be officially issued by the United Nations."

"Can all this be accomplished in time before the New Year starts, Atessa?"

"Yes Keyvan, thankfully because of the modern technology of today new calendars showing all holidays can be set up through computers worldwide and special Holiday Calendar Apps can be made available to everyone either through their Smart Phones or I Phones and for those people who don't have access to a smart phone or a computer, calendars showing the holidays on them can be issued by the governments of every country. "There will not be a Holidoomsday," Atessa emphatically says!

It was now 12 midnight on New Year's Eve, and bells chimed while fireworks illuminated the skies around the world as the New Year's holiday was ushered in and celebrated everywhere, and although Santa Claus is not normally associated or seen on this holiday, this new year was an exception. The newspapers of all countries had basically the same headline, SANTA CLAUS and the AK DETECTIVE AGENCY PREVENT HOLIDOOMSDAY. In the same newspapers, a picture was shown of the evil Xmas Repeal and his co-conspirator, Mr. Monthday, being arrested by the international Holiday Spirit Police Department, handcuffed, and taken to a secret prison to be locked up forever.

Once again, the natural order of annual holidays had been restored, as witnessed by Baby New Year's presence at the New Year's celebrations, as Old Father Time wished him well, as he was now responsible for the next 365 days of the year. While all of this was happening, Cupid received his shipment of love arrows for the Valentine Holiday that was about a month away, and Clancy, the lucky leprechaun, showed St. Patrick the shipment of four-leaf clovers and shillelaghs that had just been delivered.

As Santa Claus had told Atessa at the meeting of the World Council of Holidays just a few short days ago, she and her assistant Keyvan, who was also her brother, should be mentioned, and the AK Detective Agency will forever become legendary whenever holidays and birthdays are celebrated. In fact, I think there will someday be a new holiday celebrated each year between Christmas and New Year's. I think they will call it Holiday Savings Day. What do you think?

The End

Grandpa Ralph

www.ingramcontent.com/pod-product-compliance
Lightning Source LLC
LaVergne TN
LVHW022226200926
848481LV00006B/2145